Climate Life

Policy, Politicians, & Propaganda

By Von Knowledge

CONTENTS

Preface

The book *Climate Life* is a manifestation of my frustration with trying to understand the facts about climate change and truly getting to the bottom of the "climate crisis" as so many climate scientists are calling it these days.

No one can refute we have only one planet, at some point all water ends up being recycled, and climate has been changing since the beginning of time. The world ecosystem is in constant flux and there is nothing we can do to change this except try to better understand climate before enacting additional policy which could prove detrimental to the environment and our freedoms.

An excerpt from a 1970's article on global cooling went as far to mention *"...melting the arctic ice cap by covering it with black soot or diverting arctic rivers"*. Does that sound like a sane solution to you? I'll let you be the judge... especially since an article in 2014 stated a climate scientist, *"correctly predicted back in 2000 that the Earth was entering a cooling phase."* What are we to believe?

Within this book you will find over 100 various excerpts from research articles, science journals, magazines, newspapers, and online media sources which will hopefully help you better understand what we are up against. You will also find timely quotes which may shed some light on the political ideology behind the messaging.

Shortly after starting the process of organizing the articles and quotes for this project I quickly realized this is the type of book the Nazi's, National *Socialists*, would have chosen to burn during their 1930's book burning campaign. The quotes mentioned in this book clearly challenge the current global agenda to use climate as a way to fund socialist propaganda within colleges, redistribute wealth, remold society, and collapse Capitalism as we know it today.

However you interpret the excerpts within these pages...

"One thing is for certain, we are uncertain in our certainty."
Von Knowledge

About the Author

Born: June 1st 2017
City of Birth: None of your business
Eye Color: Seriously?
Status: Single... not interested in mingling.

If you haven't put two and two together yet Von Knowledge is an alias. Why would I use an alias for such an important book such as *Climate Life*? Personally I do not trust the biased media, socialist anarchists and/or the global elites who say they are for the little guy but choose to line their pockets while they continue flying around the globe on their private jets going from climate conference to climate conference.

To calm the savage beasts who might try to create a conspiracy theory around this book I'd like to get some facts straight. I've never worked for a power company, mining company or any other energy based company. I'm not a lobbyist nor have I ever been one, a scientist, a politician, political aid, political action committee member, a professor, a racist, a fascist, a socialist, a sexist, a homophobe, a transgenderphobe, or a xenophobe.

I AM... Von Knowledge, a 38 year old concerned citizen of the United States of America

CHAPTER 1

Constant Change

"The only thing that is constant is change." - Heraclitus

The article below was provided by the author of *The Daily Philosopher* and elaborates on one of the most important aspects of life... change. I'd personally like to thank the author for sharing his thoughts.

"The change that occurs in a river is vivid and unmistakable. By claiming that the change we see in a river is true of our world in general, Heraclitus challenges the idea that some things simply stay the same: we may not see the change so clearly, but change is occurring nonetheless. This might be easiest to accept in the physical realm, where, for example, on the level of atoms, there is constant motion in all physical objects, no matter how solid and stationary they may seem. And certainly it is easy enough to see that the bodies of all living things are constantly changing, not only aging but also going through various biological processes and exchanges with the environment, such as breathing. But what about other realms? Heraclitus might not have been thinking about things such as relationships and love, or a person's identity, but his insistence on the fundamental fact of change encourages us to consider whether change is not inevitable in such aspects of life as well.

If we take Heraclitus's model of the world as a guide, change is not only something we must accept, but it is actually something to celebrate. Heraclitus saw the world as a system in flux, but in his view that very flux is also what keeps the world the same, in a sense. In a famous re-statement, Plato leaves out that aspect of Heraclitus's view: 'Heraclitus, you know, says that everything moves on and that nothing is at rest; and, comparing existing things to the flow of a river, he says that you could not step into the same river twice.'1 So, according to Plato's way of stating the idea, the river itself is a different river from moment to moment, since the water flowing in it is different: if you step into a river at one moment and step out, and then step back in, you are stepping into a different river. But if we look carefully at the fragment from Heraclitus, we see that although he says the waters are changing, he does not say that the river is different. Heraclitus specifically claims that it is the same river although its waters are constantly changing.2 So according to Heraclitus there can be an overall stability despite, or perhaps because of, constant change: The river is the same river although it is changing--it's just part of what it is to be a river that there is this constant change going on.

Even if we do not agree with Heraclitus that all things are like this, still we may find in many realms of our lives that the only way some things can exist is by changing. A child, for example, is something that we all accept and enjoy as a constantly changing thing. A child, like a river, doesn't become a different child with each change. Being a child simply involves changing all the time. A certain kind of change seems to be a part of the basic nature of some, if not all, things.

Heraclitus's insistence on the process of change as fundamental to the world poses a question to us when we are facing difficult changes that we might want to deny or resist. By insisting that something or someone stay the same, could it be that we are destroying the very thing we wish to preserve? In any particular case, when we are resisting change, we might ask ourselves, is this like trying to stop a river's waters from flowing?"
The Daily Philosopher, August 2004
The Only thing Constant is Change

1 Trans. John Mansley Robinson, An Introduction to Early Greek Philosophy, (Boston: Houghton Mifflin Company, 1968) p. 90 (emphasis added).
2 Heraclitus refers to "the same rivers," in the plural.

CHAPTER 2

Climate Definitions

Definition of *climate change*

　　1.　: changes in the Earth's weather patterns

Definition of *global cooling*

Words fail us says Merriam-Webster online. Think we know what global cooling means. Brrrrrr

Global cooling was a conjecture during the 1970s of imminent cooling of the Earth's surface and atmosphere culminating in a period of extensive glaciation.

Definition of *global warming*

　　1.　: an increase in the earth's atmospheric and oceanic temperatures widely predicted to occur due to an increase in the greenhouse effect resulting especially from pollution

Definition of *scientific method*

　　1.　: principles and procedures for the systematic pursuit of knowledge involving the recognition and formulation of a problem, the collection of data through observation and experiment, and the formulation and testing of hypotheses

CHAPTER 3

Climate Tactics Defined

Definition of *climate alarmist*

1. : someone who rushes to judgment on a scientific finding regarding climate and spreads doomsday prophecy without all the data or facts to back up the claims

Definition of *climate denier*

1. : someone who refuses to accept human activities are contributing significantly to climate change and global warming

Definition of *gas lighting*

1. : is a form of manipulation that seeks to sow seeds of doubt in a targeted individual or members of a group, hoping to make targets question their own memory, perception, and sanity. Using persistent denial, misdirection, contradiction, and lying, it attempts to destabilize the target and delegitimize the target's belief

CHAPTER 4

1970s Climate - Global Cooling

"Climate Change moves in irregular cycles."
National Geographic, November 1976
What's Happening to our Climate

"We will be forced to sacrifice democracy by the laws that will protect us from further pollution."
Dr. Arnold Reitze,
The Argus-Press, January 26th 1970
Pollution Prospect A Chilling One

Reitze Suggested in the article:

1. Outlawing the internal combustion engine for vehicles and outlawing or strict controls over all forms of combustion.

2. Rigid controls on the marketing of new products, which will be required to prove a minimum pollution potential.

3. Controls on all research and development to be halted at the slightest prospect of pollution.

4. Possibly even population controls, the number of children per family prescribed and punishment for exceeding the limit.

"The world could be as little as 50 or 60 years away from a disastrous new ice age."
Dr. S. I. Rasool of NASA and Columbia University
Washington Post, July 9th 1971

"Man may even be able to change the climate of the earth. This is one of the most important questions of our time."
National Science Board, 1972

"Judging from the record of the past interglacial ages, the present time of high temperatures should be drawing to an end... leading into the next glacial age."
National Science Board, 1972

"During the last 20 to 30 years, world temperature has fallen, irregularly at first but more sharply over the last decade. U.S."
National Science Board, 1974

"We live in an unusual epoch: today the Polar Regions have large ice caps, whereas during most of the earth's history the poles have been ice-free."
National Academy of Sciences, 1975
US National Academy of Sciences/National Research Council Report
UNDERSTANDING CLIMATE CHANGE: A program for action

"'Our knowledge of the mechanisms of climatic change is at least as fragmentary as our data,' concedes the National Academy of Sciences report. 'Not only are the basic scientific questions largely unanswered, but in many cases we do not yet know enough to pose the key questions.'"
National Academy of Sciences, 1975
US National Academy of Sciences/National Research Council Report
UNDERSTANDING CLIMATE CHANGE: A program for action

"Climatic change has been a subject of intellectual interest for many years. However, there are now more compelling reasons for its study: the growing awareness that our economic and social stability is profoundly influenced by climate and that man's activities themselves may be capable of influencing the climate in possibly undesirable ways. The climates of the earth have always been changing, and they will doubtless continue to do so in the future. How large these future changes will be, and where and how rapidly they will occur, we do not know."
National Academy of Sciences, 1975
US National Academy of Sciences/National Research Council Report
UNDERSTANDING CLIMATE CHANGE: A program for action

"A major climate change would force economic and social adjustments on a worldwide scale."
National Academy of Sciences, 1975
US National Academy of Sciences/National Research Council Report
UNDERSTANDING CLIMATE CHANGE: A program for action

"There are ominous signs that the earth's weather patterns have begun to change dramatically and that these changes may portend a drastic decline in food production – with serious political implications for just about every nation on earth. The drop in food output could begin quite soon, perhaps only ten years from now."
Newsweek, April 28th 1975
Article: The Cooling World

"Climatologists are pessimistic that political leaders will take any positive action to compensate for the climatic change, or even to allay its effects. They concede that some of the more spectacular solutions proposed, such as melting the arctic ice cap by covering it with black soot or diverting arctic rivers, might create problems far greater than those they solve."
Newsweek, April 28th 1975
The Cooling World

"The evidence in support of these predictions has now begun to accumulate so massively that meteorologists are hard pressed to keep up with it."
Newsweek, April 28th 1975
The Cooling World

"It may seem there are many theories on climate as there are climatologists, but experts agree on one point: They cannot yet predict climate change with any assurance."
National Geographic, November 1976
What's Happening to our Climate

"Climate Change moves in irregular cycles."
National Geographic, November 1976
What's Happening to our Climate

"The scientists and computers at the National Oceanic and Atmospheric Administration were confidently predicting that the frigid weather would continue. The chilling pronouncement of NOAA'S senior climatologist: 'The forecast is for no change."
Time Magazine, January 31st 1977
The BIG Freeze

"If the arctic region continues its cooling trend, it could trigger a new ice age, with the glaciers covering a third of the Country..."
Steven S Ross,
New York Magazine, January 31st 1977

"'It doesn't look good, not in our lifetime, and it's going to be even worse for future generations,' said Madeleine Briskin of the University of Cincinnati, who specializes in researching long-range weather cycles."
Ellensburg Daily Record, January 10th 1978
Winters will get Colder, we're entering little Ice Age

"The observations come, at a time when a warming trend could have been expected from the increase of carbon dioxide in the atmosphere due to extensive fuel burning. The gas inhibits the escape of solar heat from the earth. Dr. Kukla, in a telephone interview this week, said that the cause of the apparent cooling remained unknown and that no scientific attempt to predict whether the trend would continue was possible."
Walter Sullivan,
The New York Times, January 5th 1978
International Team of Specialists Finds No End in Sight to 30-Year
Cooling Trend in Northern Hemisphere

"Dr. Lenoa M. Libby and Dr. Louis J Pandolfi of Los Angeles forecast continued bitterly cold winters all over the globe past the mid-1980's. But late next decade, they say, there will be a gradual warming trend which will continue until century's end. Then will follow a cold snap which could last through the first half of the 21st century."
Peter Smark,
The Age, January 16 1979
The New Ice Age Cometh

"As if people didn't have enough to be concerned about, Isaac Asimov has authored a book called A Choice of Catastrophes that probably will initiate a run on the worry bead market. ...the author suggests that men in the future may even use giant mirrors placed in near space, to reflect sunlight that ordinarily would miss the earth onto the earth during the cooling phases and reflect rays away from the earth if ice-melting threatens."
Spokane Daily Chronicle, October 12th 1979
Get Ready to Freeze

CHAPTER 5

1980s Climate
Global Cooling & Warming

*"Since others have explained my theory,
I can no longer understand it myself."* - Albert Einstein

"The time-dependent increase in atmospheric CO_2, from 1860 to 2025, is taken from carbon-cycle models. The model results suggest that ocean heat capacity will produce a lag in CO_2 induced global warming of about 2 decades."
Robert D. Cess and Steven D. Goldenberg
Journal of Geophysical Research, January 20th 1981
The effect of ocean heat capacity upon global warming due to increasing atmospheric carbon dioxide

"Because of global heating attributed to an increase in atmospheric carbon dioxide from fuel burning, about 20,000 cubic miles of polar ice has melted in the past 40 years, apparently contributing to a rise in sea levels, according to calculations by two scientists of the National Oceanic and Atmospheric Administration.

Although melting ice and rising sea levels seem firmly related, the researchers cautioned that the relationship might prove misleading. The 'apparent consistency,' they reported in the Jan. 15 issue of the journal Science, 'may be spurious' because of difficulty in getting reliable data."
Walter Sullivan,
The New York Times, January 8th 1982
SOME POLAR ICE MELTING LINKED TO GLOBAL HEATING

"Tonight President Reagan's science adviser, George A. Keyworth 3d, sharply criticized the E.P.A. report and praised the National Academy's. He called the environmental agency's report 'unwarranted and unnecessarily alarmist.' ...

'There is no evidence to indicate that the gradual rise in carbon dioxide in the air would have environmental effects pronounced enough to require near-term corrective action,' Mr. Keyworth said. ...

He contrasted the E.P.A. report with the academy's study, which he said 'emphasized that, at this time, there are no actions recommended other than continued research on this issue.' ...

Both reports said there were many uncertainties involved with the projections."
Philip Shabecoff,
The New York Times, October 2st 1983
HASTE OF GLOBAL WARMING TREND OPPOSED

"Despite the much heralded 'greenhouse effect' that is said to be warming the Earth's climate average temperature in Illinois have fallen significantly in the last five decades, an analysis by state weather scientists shows."
Jon Van and Eddy McNeil,
Chicago Tribune, April 23rd 1984
Cold facts suggest a chilling trend

"A series of General Circulation Model experiments are performed to examine the role of paleogeography as an explanation of the Tertiary global cooling trend.

Either a series of events, some geographically related, or another forcing factor, most probably atmospheric CO_2 concentration, are offered as alternative explanations of the Tertiary global cooling trend."
Eric J. Barron,
National Center for Atmospheric Research, 1985
Palaeogeography, Palaeoclimatology, Palaeoecology

Explanations of the tertiary global cooling trend

"Natural climate variations are masking this temperature increase, but further additions of trace gases during the next 65 years could double or even quadruple the present effects, causing the global average temperature to rise by at least 1 °C and possibly by more than 5 °C."
Robert E. Dickinson & Ralph J. Cicerone,
Nature Publishing Group, January 9th 1986
Future global warming from atmospheric trace gases

"The frigid air over Antarctica took three weeks longer than usual to warm at the onset of the Antarctic spring this year, prompting concern that the 'ozone hole' discovered over the continent less than three years ago may be affecting global climate. ...

University of California scientist F. Sherwood Rowland, a leading expert in ozone depletion, said the event 'could be the first indication of major climate change. There is no way of judging the impact, but it's an ominous trend."
Milwaukee Sentinel - Washington Post Cited, December 21st 1987
NOAA 1987 : Climate Change Making Antarctica Colder

"The earth has been warmer in the first five months of this year than in any comparable period since measurements began 130 years ago, and the higher temperatures can now be attributed to a long-expected global warming trend linked to pollution, a space agency scientist reported today.

Dr. Hansen, a leading expert on climate change, said in an interview that there was no 'magic number' that showed when the greenhouse effect was actually starting to cause changes in climate and weather. But he added, 'It is time to stop waffling so much and say that the evidence is pretty strong that the greenhouse effect is here.' An Impact Lasting Centuries.

If Dr. Hansen and other scientists are correct, then humans, by burning of fossil fuels and other activities, have altered the global climate in a manner that will affect life on earth for centuries to come."
PHILIP SHABECOFF,
The New York Times, June 24th 1988
Global Warming Has Begun, Expert Tells Senate

"Title XI: World Population Growth -Declares it is the policy of the United States that family planning services should be made available to all persons requesting them. Authorizes appropriations for FY 1991 through 1995 for international population and family planning assistance. Prohibits the use of such funds for: (1) involuntary sterilization or abortion; or (2) the coercion of any person to accept family planning services.

Requests the President to initiate an international conference on population, and to seek an international agreement on population growth. Establishes a National Commission on Population, Environment, and Natural Resources to prepare reports and convene conferences. Terminates such Commission three years after the enactment of this Act.

Mandates that multilateral development banks adopt guidelines promoting lending strategies which emphasize the maintenance of sustainable world population levels. Authorizes appropriations for FY 1991 through 1993."
101st U.S. Congress February 22th 1989
H.R.1078 - Global Warming Prevention Act of 1989

"Now, as the 1990s approach, the talk has turned to the science of survival--saving forests for oxygen, keeping streams from spreading toxic pollutants, cleaning the air to avoid catastrophic global warming. ...

And the Worldwatch Institute, an environmental research organization, calls the '90s 'the turnaround decade' in which people will either stop polluting or face an environmental disaster as devastating as nuclear war.

'By many measures, time is running out,' Worldwatch warned in its 'State of the World 1989' report.

Not everyone shares Worldwatch's apocalyptic vision, which is based largely on the threat of global warming--the 'greenhouse effect.' Many respected scientists say the available evidence does not warrant the doomsday warnings."
Mitchell Landsberg
Associated Press – Los Angeles Times, October 29th 1989

"The skeptics are properly credentialed, and the gist of their comments is that the greenhouse effect is not a 100 percent certainty. The computer models on which projections have been based are flawed, they argue, and the climatological data presented as evidence may actually be the reflection of long-term natural cycles that are not fully understood (and thus could benefit from some more research grants).

Their comments are catching the ears of policy makers because if there is no global warming, then all the costly preparations being suggested to avert it, or to minimize its effects, will be not only unnecessary but also counterproductive."
Boston Globe, December 17th 1989
COMMON SENSE AND GLOBAL WARMING

CHAPTER 6

1990s Climate - Global Warming

"There are no facts, only interpretations." - Friedrich Nietzsche

"Some important priorities may be defined. First, reliable projections of human settlement implications of climate change should relate to specific climate models, none of which can yet provide reliable projections of likely future local climates. ...

Third, the relationship between urban, social and economic changes and climatic effects needs to be quantified."
Intergovernmental Panel on Climate Change (IPCC), 1990
Potential impacts of climate change

"The gradual depletion of the ozone layer and the related 'greenhouse effect' has now reached crisis proportions as a consequence of industrial growth, massive urban concentrations and vastly increased energy needs. Industrial waste, the burning of fossil fuels, unrestricted deforestation, the use of certain types of herbicides, coolants and propellants: all of these are known to harm the atmosphere and environment. The resulting meteorological and atmospheric changes range from damage to health to the possible future submersion of low-lying lands."
Saint John Paul II,
GLOBAL CATHOLIC CLIMATE MOVEMENT, 1990
Message for the World day of Peace, "Peace with God the Creator, Peace with all of creation"

"The assessment of the proper response to the possible danger of global warming depends critically on the determination of how real the danger is. ...

The existence of skepticism on this issue has only recently been publicly recognized. Whatever the truth may turn out to be, there is an unusual degree of extremism associated with this issue. While environmental scares are not unheard of, few have been accompanied by recommendations that skepticism be stifled (an editorial to this effect in the Boston Globe [17 December 1989] is but one of a series of examples). As an admitted skeptic on this issue, I would like to discuss some aspects of the 'greenhouse hypothesis' that leaves me unconvinced, and leave me concerned whether unanimity on such an issue is healthy for meteorology. ...

The current state of our understanding of climate hardly justifies a consensus over the response of climate to the small increase in downward flux caused by a doubling of CO_2. It is not clear that models will ever be able to deal with this issue with great certainty."
Richard S. Lindzen
Center for Meteorology and Physical Meterology, March 1990
MIT, Cambridge, MA, USA
Some Coolness Concerning Global Warming

" ... These uncertainties are forcing the international community to make difficult decisions regarding responses needed to counter the looming threat of global warming. The choice appears to be between adopting a precautionary approach today, one that will be very expensive and that may itself alter society in fundamental ways, or waiting for the results of the current 'experiment,' and suffer the cataclysm that could result from making the wrong guess-a planet warming so rapidly that life may not be able to adapt. ...

The choice seems clear: international society does not have the luxury of waiting for scientific certainty before it responds to the potential threat of global warming. The international community must construct a global precautionary agreement to first, reduce emission of greenhouse gases to

a safe level and second, to ensure that future development becomes sustainable.8 This requires the participation and cooperation of all elements of international society working together to achieve such an agreement fast enough to avoid the most severe impacts of global warming, including sea-level rise."
Durwood Zaelke & James Cameron
American University International Law Review, 1990
Global Warming and Climate Change – An Overview of the International Legal Process

"A report issued by the U.S. space agency NASA concluded that there has been no sign that the greenhouse effect increased the global temperature in the 1980s. Based on satellite analysis of the atmosphere between 1,500 and 6,000 meters above sea level, the report said that the study found 'a seemingly random pattern of change from year to year.'"
Steve Newman,
The Canberra Times, April 1st 1990
Earthweek: A Diary of the Planet

"In the last 20 years, eminent scientists continued to ridicule the theory of continental drift. The theory of global climate change used to be ridiculed, too. But in the last few years, the overwhelming majority of scientists who have examined the evidence have agreed that the problem is real. ...

People are changing their thinking about the importance of protecting the global environment. We too are showing our willingness to act. The obstacles may seem immovable, but so did the Berlin wall. With bold leadership and a new political 'ecolibrium,' we too shall overcome."
Al Gore Jr, Democrat, Senator from Tennessee,
The New York Times, April 22nd 1990
To Skeptics on Global Warming . . .

"The public policy dilemma is what action to take even though we will not know in detail what will happen-the scientific community will not be able to provide much definitive information over the next decade or two about

the precise timing and magnitude of century-long climate changes, especially if research efforts remain at current levels. ...

Ideology is the enemy of flexibility. It is incumbent on all of us to rethink any of our ideologically rigid positions in order to fashion ways in which to enhance the flexibility of management."
Stephen H. Schneider,
National Center for Atmospheric Research, September 1990
The Global Warming Debate Heats Up: An Analysis and perspective

"Climatologists readily admit that they cannot predict with any certainty what the meteorological effects of climate change will be for a particular area. ...

The response of the international community, individually and collectively, to the possibility of global climate change has been both to seek to prevent such change and, additionally, to prepare to adapt to the changes that are likely to occur despite such prevention efforts. The scholarly agenda regarding both prevention and adaptation strategies to climate change is growing but is not yet clearly articulated for many disciplines, including law. This Article deals with an aspect of adaptation to climate change: the effect of a rising sea level on the law of baselines. ...

Until late 1989, a conservative and widely accepted estimate advanced in, among other places, a 1987 National Research Council study was that a doubling in the amount of carbon dioxide present in the atmosphere in 1950 (estimated to occur between 2010 and 2050), would cause the sea to rise 1 meter, or 3.3 feet. In 1989, an emerging view held that this estimate was too high. Mark Meier, Director of the Institute of Arctic and Alpine Research at the University of Colorado estimated a one-third-meter rise in sea level, but added that the margin of error was 'horrendous.'"
David D. Caron,
Ecology Law Quarterly, September 1990
When Law Makes Climate Change Worse: Rethinking the Law of Baselines in Light of a Rising Sea Level

"We have become more and more aware of the growing imbalance between our species and other species, between population and resources, between humankind and the natural order of which we are part. ...

The real dangers arise because climate change is combined with other problems of our age: for instance the population explosion; — the deterioration of soil fertility; — increasing pollution of the sea; — intensive use of fossil fuel; — and destruction of the world's forests, particularly those in the tropics."
Margaret Thatcher
2nd World Climate Conference, November 6th, 1990

"The IPCC report is a remarkable achievement. It is almost as difficult to get a large number of distinguished scientists to agree, as it is to get agreement from a group of politicians. As a scientist who became a politician, I am perhaps particularly qualified to make that observation! I know both worlds."
Margaret Thatcher
2nd World Climate Conference, November 6th, 1990

"Laboratory studies of the effects of elevated CO_2 levels on plants have documented increased rates of photosynthesis, lowered plant water use requirements, increased carbon sequestering and increased soil microbial activity fixing nitrogen for fertilizer, thereby stimulating growth (Hardy and Havelka, 1975; Drake et al., 1988). Therefore, CO_2 increases could theoretically provide significant benefits for plants and trees undergoing water stress in drier climates. ...

The concept of the greenhouse effect has been widely if not universally accepted. However, because of the still-crude capabilities of current global circulation models to model complex terrestrial-ocean-atmospheric interactions, debate remains about how much future climates are likely to warm, and when."
Kenneth Andrasko,
Food and Agriculture Organization of the United Nations, 1990

Global warming and forests: An overview of current knowledge

"Of course, much more research is needed. We don't yet know all the answers. Some major uncertainties and doubts remain. No-one can yet say with certainty that it is human activities which have caused the apparent increase in global average temperatures. The IPCC report is very careful on this point. ...

Nor do we know with any precision the extent of the likely warming in the next century, nor what the regional effects will be, and we can't be sure of the role of the clouds. ...

Global climate change within limits need not by itself pose serious problems—our globe has after all seen a great deal of climate change over the centuries. And it's notable that the blue-green algae which dominated the Precambrian period at the dawn of life are still major components of the marine phytoplankton today. Despite the climate changes of many millions of years, these microbes have persisted on earth virtually unchanged, pumping out life-giving oxygen into the atmosphere and mopping up carbon dioxide.

Britain will continue to play a leading role in trying to answer the remaining questions, and to advance our state of knowledge of climate change. ...

We need to improve in particular our understanding of the effect of the oceans on our weather, improve too our capability to model climate change."
Margaret Thatcher
2nd World Climate Conference, November 6th, 1990

"The lessons from the past tell a chilling tale about the warming of the future, according to paleontologists and anthropologists at the Smithsonian Institution. And now they too have joined in the debate about global change. Concerned that their institution, the National Museum of Natural History, has been overlooked by the nation's global change policymakers, they have exchanged lab coats for suit jackets and lab benches for podiums to let Washington know they want a larger role in the government's $1 billion research program.

'In some of these areas, I don't think there's anyone in the government that is producing the kind of long-term data we need,' says Daniel Appleman, the museum's associate director for science. 'Our scientists want to be part of this effort.'"
Elizabeth Pennisi,
The Scientist, May 1990
Fossil Record Aids In Predictions Of Global Warming's Consequences

"The idea that developing countries like India and China must share the blame for heating up the earth and destabilizing its climate, as espoused in a recent study published in the United States by the World Resources Institute in collaboration with the United Nations, is an excellent example of environmental colonialism. ...

The report of the World Resources Institute (WRI), a Washington- based private research group, is based less on science and more on politically motivated and mathematical jugglery (1). Its main intention seems to be to blame developing countries for global warming and perpetuate the current global inequality in the use of the earth's environment and its resources. ...

Those who talk about global warming should concentrate on what ought to be done at home. The challenge for India is thus to get on with the job at hand and leave the business of dirty tricks and dirtying up the world to others. In this process, we will help ourselves and may be even, the rest of the world."
 Anil Agarwal and Sunita Narain,

Centre for Science and Environment, New Delhi, 1991
GLOBAL WARMING IN AN UNEQUAL WORLD, a case of environmental colonialism

"We hypothesis that the global and hemispheric temperature series are the result of a Markov process. The climate system is subjected to various forms of random impulses. It is argued that the system fails to return to its former state after reacting to an impulse but tends to adjust to a new state of equilibrium as prescribed by the shock. This happens because a net positive feedback accompanies each shock and slightly alters the environmental state.

It is important to examine all ways and means by which the observed data series develop trends before facing hand and fast conclusions that any particular activity is the one and only responsible agent."
A.H. Gordon,
Flinders Institute for Atmospheric and Marine Science, SA, June 1991
Global Warming as a Manifestation of a Random Walk

"This summit, formally called the United Nations Conference on Environment and Development, or UNCED, will discuss issues ranging from the distribution of wealth among nations and women's rights to deforestation. But the topic that will attract the most attention and controversy is the claim that the Earth is subject to steady and potentially damaging rise in temperature -- a phenomenon known as "global warming" -- and that this condition is in large part a byproduct of Western industrial growth."
John Shanahan,
The Heritage Foundation, May 21st 1992
A Guide to the Global Warming Theory

"In spite of these fears, the accumulated scientific data do not support such dire predictions, showing the cataclysmic results to be either highly improbable or simply wrong. Moreover, there is enormous uncertainty

associated with the scientific methodology used to predict future climate changes. Among the difficulties:

Climate change computer models that predict warming often rely on assumptions and simplifications that raise questions about their reliability."
John Shanahan,
The Heritage Foundation, May 21st 1992
A Guide to the Global Warming Theory

"Despite the scientific consensus on projected global-scale changes in temperature, many uncertainties remain, particularly regarding regional-scale changes in precipitation and other climatic elements. Five years from now, we may not be much closer to reducing these uncertainties."
S.J. Cohen,
Canadian Water Resources Journal, 1992
IMPACTS OF GLOBAL WARMING IN AN ARCTIC
WATERSHED

"Given the normal climate variability, we may reasonable expect there to be future climates both warmer and colder than the present regime. This, however, hardly supports the current fear that increasing greenhouse gases in the atmosphere will lead to catastrophic warming."
Richard S. Lindzen
Center for Meteorology and Physical Oceanography, 1993
MIT, Cambridge, MA, USA
On the Scientific Basis for Global Warming Scenarios

"Existing climatic prediction models are too coarse for the development of reliable and credible regional climatic scenarios. There is a significant uncertainty regarding the climate change scenarios for sub-Saharan Africa with conflicting scenarios about which areas will get wetter and which will get drier."
Joseph H. Kinuthia
Kenya Meteorological Department, Nairobi, Kenya, 1993

Global Warming and Climate Impacts in Southern Africa: How Might Things Change?

"Although the first signals of greenhouse warming have not been detected to date, either through measurements of temperature or impacts on forest ecosystems, and concrete actions are unlikely to be undertaken until positive proof exists, it would be wise to begin investigating policy options at this time."
Brian J. Stocks
The Forestry Chronicle, June 1993
Global warming and forest fires in Canada

"While ecologists involved in management or policy often are advise to learn to deal with uncertainty, there are a number of components of global environmental change of which we are certain-certain that they are going on, and certain that they are human caused. ...

We cannot prevent global change. However, what we do can make a difference-individually by affecting particular human activities, collectively by helping to create policies and world-views that will affect which pathways of population growth and resource use human society follows. It is up to us."
Peter M. Vitousek,
Department of Biological Sciences, Stanford University, August 1993
Beyond Global Warming: Ecology and Global Change

"In summary, record low ozone levels have been observed in recent years, and substantially larger future global depletions in ozone would have been highly likely without reductions in human emissions of ozone-depleting gases. However, worldwide compliance with current international agreements is rapidly reducing the yearly emissions of these compounds. As these emissions cease, the ozone layer will gradually improve over the next several decades. The recovery of the ozone layer will be gradual because of the long times required for CFCs to be removed from the atmosphere."
National Oceanic and Atmospheric Administration (NOAA), 1994
Scientific Assessment of Ozone Depletion

"Some words of caution are necessary at the end of this chapter. The effect of non-linearities in the aerosol effects on direct and cloud-amplified forcing have already been pointed out above. These effects are superimposed on the climate system, which itself is described by a system of coupled non-linear differential equations. The chaotic behaviour characteristic of such systems results in unpredictable fluctuations at all time-scales and a tendency to jump between highly disparate states of the system, e.g. ice-ages and interglacials, or different modes of ocean circulation (BROECKER et al., 1985; BROECKER, 1987; Chapter 14 by PENG, 1994). The tendency for the Earth's climate system to undergo fast jumps at time-scales of decades to centuries has been recently documented in ice cores obtained from Greenland (GRIP (GREENLAND ICE-CORE PROJECT) MEMBERS, 1993). This study suggests that the stable climate of the present interglacial is a rather unusual case, and that previous interglacials were characterized by frequent, short-term oscillations between considerably warmer and colder states. Temperature changes of up to 10~ in a couple of decades appear to be possible in interglacials, making human and ecological adaptation to climate change highly unlikely."

Andreae Meinrato,
OAI, 1995
Climatic effects of changing atmospheric aerosol levels

"There is no global long-term trend in any rainfall change over the period of instrumental record (c. 150 years), but there has been an increase of 0·5°C in global temperature over the past 100 years. This increase seems partly due to urbanization, as there is no evidence of it resulting from atmospheric pollution by CO_2 and other warming gases (SO_2, NO_2, CH_4, CFH etc.)."

Henry N. Le Houérou,
Journal of Arid Environments, 1996
Climate change, drought and desertification

"Why has the globe warmed? Because we are confident that human activities have substantially changed the atmospheric composition in terms of greenhouse gases (GHGs; especially carbon dioxide) and aerosols, we are also confident that at least part of the observed warming is human-induced. The leading question is how much? To answer this, we first need to estimate the magnitude of the expected anthropogenic warming. To do this requires a knowledge of the anthropogenic forcing change, and a suitable model to convert this forcing to an estimated climate change. ...

... The potential importance of century time scale internal variability as a component of the observed warming has rarely been considered. Any such overall trend could be positive or negative. There is no way to determine whether such a trend component exists, let alone estimate its magnitude, but it is unlikely that it would be zero. ...

Where do we go from here? In the most recent IPCC report (31) and in an earlier report (34), it was noted that studies of global-mean temperature alone are insufficient to show a compelling cause–effect relationship between anthropogenic forcing and climate change. Such studies, as shown above, can demonstrate that the observed warming is consistent with a substantial anthropogenic effect on climate but cannot accurately quantify this effect. To show a cause–effect linkage, more sophisticated techniques are required that make use of the patterns of observed climate change, either in the nearsurface horizontal (latitudeylongitude) plane (39, 40) or in the vertical (zonal meanyheight) plane (41, 42). Such pattern based studies have shown increasing and statistically significant similarities between model predictions and observed temperature changes. These results, combined with the evidence from global-mean analyses, provide convincing evidence for a discernible human influence on global climate; but further work is required to better quantify the magnitude of the human influence and reduce uncertainties in the climate sensitivity."

T. M. L. WIGLEY*, P. D. JONES†, AND S. C. B. RAPER†
The National Academy of Sciences, August 1997
The observed global warming record: What does it tell us?

"Most Americans are willing to join other countries in setting standards to improve the global environment and a majority would even pay more for gasoline to reduce global warming. But on the eve of the December Kyoto conference on climate change, the American public strongly rejects the notion that the United States should bear more of the burden of repairing the environment than poorer countries, even when the consideration that these nations have not caused as much damage as the U.S. is raised. ...

Republicans, Democrats and Independents alike think all countries rich or poor should now share equally in global clean-up efforts. Fully 70% of Pew's respondents felt this way, compared to just 19% who believed that poorer countries should be allowed to do less. Even Americans who are aware of the fact that the United States produces more carbon dioxide per capita than other countries believe that the U.S. should not bear more of the burden. ...

Opponents of a gasoline price increase and internationally mandated standards may also take some comfort in the findings of Pew longitudinal surveys that reveal declines since 1992 in strong support for environmental regulation. This drop is consistent with less alarm about a range of environmental issues compared to the early 1990's."
Pew Research Center, November 21st 1997
Americans Support Action on Global Warming

"Within the U.S. and the rest of the Americas, there will be both winners and losers, with some areas benefitting from increases in agricultural production as a result of climate change while other areas suffer decreases. Climate change may also affect the welfare of economic groups differently (e.g. consumers vs producers). Overall, however, the consensus of economic assessments is that climate change of the magnitudes currently being discussed by IPCC and other organizations will have only a small (likely positive) effect on U.S. agriculture.

On a global scale, the regional increases and decreases associated with climate change are not expected to result in large changes in food production over the next century. Nonetheless, impacts on regional and local food supplies in some low latitude regions could amount to large

percentage changes in current production. Climate change may therefore impose significant costs on these areas. In addition, warming beyond that reflected in current studies may impose greater costs in terms of aggregate food supply. Projections from most economic studies show substantial economic losses as temperature increases beyond the equivalent of a CO2 doubling. This reinforces the need to determine the magnitude of warming which may accompany the CO2 buildup currently under way in the atmosphere."
Richard M. Adams, Brian H. Hurd, Stephanie Lenhart, Neil Leary
Department of Agricultural and Resource Economics, Oregon State University, December 17th 1998
Effects of global climate change on agriculture: an interpretative review

"The physics of climate and of climate changes associated with increasing concentrations of greenhouse gases in the atmosphere are briefly presented. Construction of a 'toy model' of the climate is discussed. Possibilities for reducing carbon dioxide emissions are indicated. Degrees of uncertainty characterizing predictions of climate responses to anthropogenic greenhouse gas emissions are presented."
John R. Barker and Marc H. Ross,
American Journal of Physics, June 1999
An introduction to global warming

"The source of carbon in biogenic methane comes from the active carbon pool. Transformation of carbon into methane, through biogenic cycling of carbon implies that the global warming (GW) contributed by biogenic methane inherits the GW of CO2. For a precise and realistic assessment of GWP of biogenic methane and its' contribution to GW, the instantaneous radiative forcing of CO2 should be subtracted from the instantaneous radiative forcing of biogenic methane. The correction suggested on this account will decrease the GWP of biogenic methane by 5%. The proposed correction is significant, since 80% of the global emission of methane involve biospheric carbon belonging to the active carbon pool."
C. K. VARSHNEY and ARUN K. ATTRI
The Wiley Online Library, July 1999
Global warming potential of biogenic methane

"Placing human well-being at the centre of concern for the environment is actually the surest way of safeguarding creation; this in fact stimulates the responsibility of the individual with regard to natural resources and their judicious use."
Saint John Paul II,
GLOBAL CATHOLIC CLIMATE MOVEMENT, 1999
Message for the World day of Peace, "Respect for human rights: the secret of true peace"

CHAPTER 7

2000s Climate - It's a Mixed Bag

"I have not failed. I've just found 10,000 ways that won't work."
- Thomas Edison

"So has air pollution gotten worse? Quite the contrary. In the most recent National Air Quality Trends report, the U.S. Environmental Protection Agency—itself created three decades ago partly as a response to Earth Day celebrations—had this to say: 'Since 1970, total U.S. population increased 29 percent, vehicle miles traveled increased 121 percent, and the gross domestic product (GDP) increased 104 percent. During this same period, notable reductions in the air quality concentrations and emissions took place.' Since 1970, ambient levels of sulfur dioxide and carbon monoxide have fallen by 75%, while total suspended particulates like smoke, soot, and dust have been cut by 50 percent since the 1950s."
Ronald Bailey,
Reason Magazine, May 2000
Earth Day, Then and Now
The planets future has never looked better. Here's why

"...because methane has a shorter atmospheric lifetime and greater radiative absorption capacity than carbon dioxide, methane reduction strategies offer an effective means of slowing global warming in the near term."
Angela Moss, Jean-Pierre Jouany, John Newbold,
INRA/EDP Sciences, January 1st 2000
Methane production by ruminants: its contribution to global warming

"One final prediction, of which I'm most absolutely certain: There will be a disproportionately influential group of doomsters predicting that the future–and the present–never looked so bleak."
Ronald Bailey,
Reason Magazine, May 2000
Earth Day, Then and Now
The planets future has never looked better. Here's why

"Recent reconstructions of Northern Hemisphere temperatures and climate forcing over the past 1000 years allow the warming of the 20th century to be placed within a historical context and various mechanisms of climate change to be tested. Comparisons of observations with simulations from an energy balance climate model indicate that as much as 41 to 64% of preanthropogenic (pre-1850) decadal-scale temperature variations was due to changes in solar irradiance and volcanism. ...

A 21st-century global warming projection far exceeds the natural variability of the past 1000 years and is greater than the best estimate of global temperature change for the last interglacial."
Thomas J. Crowley,
Science, July 14th 2000
Causes of Climate Change Over the Past 1000 Years

"The First Earth Day was the brainchild of Gaylord Nelson, the Democratic senator from Wisconsin."
Ronald Bailey,
Reason Magazine, May 2000
Earth Day, Then and Now
The planets future has never looked better. Here's why

"Among other important milestones from the cruise, scientists discovered an as yet unexplained "discontinuity" of volcanic activity along the Gakkel Ridge. Because the southern end of the ridge is spreading relatively quickly and the northern end extremely slowly, the researchers expected volcanic activity to gradually die out as they sailed north. Instead, there were irregular pockets of activity as the cruise moved northwards. ...

'We found more hydrothermal activity on this cruise than in 20 years of exploration on the mid-Atlantic Ridge,' said Charles Langmuir, co-chief scientist on Healy from Lamont-Doherty Earth Observatory (LDEO) at Columbia University."
National Science Foundation
Science Daily, November 29th 2001
Healy Researchers Make A Series Of Striking Discoveries About Arctic Ocean

"Climate prediction analysis involves complex computer modeling recipes, taking into account the buildup of greenhouse gases, the role of oceans, uncertainty about technological change, and incomplete knowledge of how the Earth's climate has changed over thousands of years.

While there is general agreement among most climate scientists that global warming is real, and that it is at least partially attributable to human activity, some climate scientists sharply disagree.

The dissenters argue that the Earth has been in a long-term, natural cycle of rising temperatures since the so-called 'Little Ice Age' 500 years ago. Further, they point out that many temperature monitoring stations are located in cities, where heat-absorbing buildings and pavement can give misleadingly high temperature readings."
CNN, April 18th 2002
Studies: Global warming to worsen

"Other human activities, such as emissions of sulfate and soot particles and the development of urban areas, have significant but more localized climate impacts. Such activities sometimes cause temperatures to rise or fall, but not by enough to offset the impact of greenhouse gases. ...

Karl and Trenberth say more research is needed to pin down both the global and regional impacts of climate change. Scientists have yet to determine the temperature impacts of increased cloud cover or how changes in the atmosphere will influence El Niño, the periodic warming of Pacific Ocean waters that affects weather patterns throughout much of the world."
National Science Foundation, December 3rd 2003
Top Scientists Conclude Human Activity Is Affecting Global Climate

"Washington state's top polluter isn't a pulp mill, a power plant or refinery. It's the newly awakened Mount St. Helens. Since the volcano began erupting in early October, it has been pumping out 50 to 250 tons a day of sulfur dioxide, the lung-stinging gas that causes acid rain and contributes to haze.

Those emissions are so high that if the volcano were a new factory, it probably couldn't get a permit, Clint Bowman, an atmospheric physicist for the Washington Department of Ecology, told The Seattle Times."
Associated Press, NBC News – Science, February 12th 2004
Mount St. Helens top Washington polluter, Up to 250 tons of sulfur dioxide a day plus haze

"Some climate scientists warn that the pace of global warming could be much more rapid than that predicted even a few years ago. ...

'Any time you get into projections, you get into a lot of uncertainties. But the [climate] models are getting a lot stronger,' said Jay Gulledge, a senior research at the Pew Center on Global Climate Change in Arlington, Virginia. ...

Gulledge cautions, however, that warming rates depend on many factors, some of which have yet to be discovered."
Brian Handwerk,
National Geographic, July 27th 2005
Global Warming: How Hot? How Soon?

"Is global warming really a threat?

Absolutely, respond most scientists, but they have only recently been able to approach a basic agreement about our changing climate. ...

John Christy, director of Earth System Science Center and critic of severe warming predictions, says forecasting the future 'gets messy quickly.'

'The Earth system has more unknowns that we are generally willing to acknowledge,' he told CNN via e-mail. 'It is very difficult for [scientists] to say, 'I don't have a clue.'...Our pronouncements often express more confidence than is warranted given the level of ignorance in which we presently operate.'

Climate models inherit this uncertainty. The first crude models -- spinning aluminum dishpans in the 1950s -- have evolved into some of the world's most sophisticated computer simulations replicating the interaction between the atmosphere, oceans and continents. Yet the system's complexity -- a mathematical swamp of biological cycles, ocean circulation, geologic emissions and even solar activity -- injects guesswork into the science."
Michael Coren,
CNN, February 10th 2006
The science debate behind climate change: Forecasting the future remains a contentious exercise

"BRITAIN'S Channel 4 has produced a devastating documentary titled 'The Great Global Warming Swindle.' It has apparently not been broadcast by any U.S. networks, but is available on the Web.

Distinguished scientists specializing in climate and climate-related fields talk in plain English and present readily understood graphs showing what a crock the current global-warming hysteria is."
Thomas Sowell
NY Post, March 17th, 2007
THE GLOBAL-WARMING SCAM

"In his program, Mr. Durkin rejects the concept of man-made climate change, calling it 'a lie ... the biggest scam of modern times.'

The truth, he says, is that global warming 'is a multibillion-dollar worldwide industry, created by fanatically anti-industrial environmentalists, supported by scientists peddling scare stories to chase funding, and propped up by compliant politicians and the media.'

Channel 4 says that the program features 'an impressive roll-call of experts,' including nine professors, who are experts in climatology, oceanography, meteorology, biogeography and paleoclimatology."
Washington Times, March 6th 2007
Global warming labeled a 'scam'

"A British judge has ruled that Al Gore's Oscar-winning film on global warming, 'An Inconvenient Truth,' contains 'nine errors.' ...

But he also said Gore makes nine statements in the film that are not supported by current mainstream scientific consensus. Teachers, Burton concluded, could show the film but must alert students to what the judge called errors. ...

The judge said that, for instance, Gore's script implies that Greenland or West Antarctica might melt in the near future, creating a sea level rise of up to 20 feet that would cause devastation from San Francisco to the Netherlands to Bangladesh. The judge called this 'distinctly alarmist' and said the consensus view is that, if indeed Greenland melted, it would release this amount of water, 'but only after, and over, millennia.'

Burton also said Gore contends that inhabitants of low-lying Pacific atolls have had to evacuate to New Zealand because of global warming. 'But there is no such evidence of any such evacuation,' the judge said.

Another error, according to the judge, is that Gore says 'a new scientific study shows that for the first time they are finding polar bears that have actually drowned swimming long distances up to 60 miles to find ice.' Burton said that perhaps in the future polar bears will drown 'by regression of pack-ice' but that the only study found on drowned polar bears attributed four deaths to a storm.

The ruling comes amid speculation that Gore will win the Nobel Peace Prize on Friday for his work on global warming."
Mary Jordan,
Washington Post Foreign Service, October 12th 2007
U.K. Judge Rules Gore's Climate Film Has 9 Errors

"Easily one of the most important stories of 2008 has been all the evidence suggesting that this may be looked back on as the year when there was a turning point in the great worldwide panic over man-made global warming. Just when politicians in Europe and America have been adopting the most costly and damaging measures politicians have ever proposed, to combat this supposed menace, the tide has turned in three significant respects.

First, all over the world, temperatures have been dropping in a way wholly unpredicted by all those computer models which have been used as the main drivers of the scare. Last winter, as temperatures plummeted, many parts of the world had snowfalls on a scale not seen for decades. This winter, with the whole of Canada and half the US under snow, looks likely to be even worse. After several years flatlining, global temperatures have dropped sharply enough to cancel out much of their net rise in the 20th century.

Ever shriller and more frantic has become the insistence of the warmists, cheered on by their army of media groupies such as the BBC, that the last 10 years have been the "hottest in history" and that the North Pole would

soon be ice-free – as the poles remain defiantly icebound and those polar bears fail to drown. All those hysterical predictions that we are seeing more droughts and hurricanes than ever before have infuriatingly failed to materialise."
Christopher Booker,
UK Telegraph, December 27th 2008
2008 was the year man-made global warming was disproved

"... The earth's surface temperature is not at record levels: According to NASA's Goddard Institute for Space Studies, surface air temperature measurements show that December 2007 to November 2008 was the coolest year, and that the hottest decade was the 1930s, not the 1990s. ...

Temperatures are still dropping: NASA satellite readings on global temperatures show that August was the fourth month in 2008 when temperatures fell below their 30-year average since satellite records began."
National Center for Policy Analysis, January 2nd 2009
GLOBAL WARMING: REASONS WHY IT MIGHT NOT ACTUALLY EXIST

"The Emissions Trading Scheme legislation poises Australia to make the biggest economic decision in its history, yet there has been no scientific due diligence.

There has never been a climate change debate in Australia. Only dogma. To demonise element number six in the periodic table is amusing. Why not promethium? Carbon dioxide is an odourless, colourless, harmless natural gas. It is plant food. Without carbon, there would be no life on Earth.

The original source of atmospheric CO2 is volcanoes. The Earth's early atmosphere had a thousand times the CO2 of today's atmosphere. This CO2 was recycled through rocks, life and the oceans."
Ian Plimer,
The Australian, May 29th 2009
Vitriolic climate in academic hothouse

"The Obama Administration on Tuesday released a report showing climate disruption is already leaving deep imprints on every sector of the environment and that the consequences of these changes will grow steadily worse in coming decades.

The report, Global Climate Change Impacts in the United States, is issued every decade by the federal government's Climate Change Science Program. It is an attempt to consolidate and transcribe into accessible language the latest climate science across a broad spectrum of disciplines and regions.

The report notes that reducing carbon dioxide emissions could lessen warming this century and beyond. But it makes equally clear that climate-related changes are already being observed globally and that new problems and challenges will develop no matter how radically emissions are reduced in the future."
Douglas Fischer,
Scientific American, June 17, 2009
Global Warming Impacts In Every Corner of the United States

"A leading UK lawyer, who represented the parent that sued Al Gore in the British High Court, has laughed off claims by the former vice-president that the judge ruled in his favour. Speaking from London John Day, a senior partner in Malletts Solicitors, said Mr Gore was misrepresenting what the judge had found. Mr Day represented a British parent who sued the UK Ministry of Education when they wanted to distribute and show Mr Gore's documentary An Inconvenient Truth to every British school child. In the 2006 documentary Mr Gore claimed humanity is in danger because of man made Global Warming. He also claimed flooding and disease would increase with the destruction of most of the world's major cities including New York, London and Shanghai. As a result Mr Gore was awarded a Nobel Peace Prize and the documentary won an Oscar.

However, after a lengthy hearing a High Court Judge, Mr Justice Burton, found that An Inconvenient Truth contained significant scientific errors in nine key areas. But questioned about the embarasing High Court decision

during a current trip to Australia Mr Gore stated on ABC Australia 'Well, the ruling was in my favour.' However, this has been rejected by Mr Day who said Mr Gore's latest claims are 'difficult to square with the reality of the judgement'. 'The judge found there were nine serious scientific errors in the film.' He said the court ordered that the film was 'not suitable to be shown in British schools without a health warning'. 'Mr Justice Burton said an Inconvenient Truth wasn't fit to be shown in British schools without suitably corrected guidance which drew attention to the errors in the film and its political partisanship.' Among the errors listed by Mr Justice Burton were Mr Gore claims that rising sea levels would destroy cities in the near future, that the polar bear was endangered and that the snows of Kilimanjaro were melting all because of Global Warming. The judge found these to be scientific errors. He also dismissed Mr Gore's claims that Hurricane Katrina was caused by Global Warming."
Ann McElhinney & Phelim McAleer,
The Climate Depot, July 19th 2009
UK Lawyer Slams Gore Over Court Case Claims

"Capitalism, in which global inequality is the norm, also avoids real democracy. This is why, to overcome climate change, we have to transform society into one based on democracy and economic equality. Central to this is expanding Aboriginal land rights, poverty reduction, refugee rights, gender equality and workers' rights. ...

There is already too much carbon in the atmosphere. The warming already in the system risks the crossing of various natural 'tipping points' that would change the earth's life support systems irreversibly on human timescales, and could raise temperatures further and faster. ...

We have enough sun and wind to provide all our energy many times over. We also have the wealth to develop a renewable energy manufacturing industry and other appropriate technology. We can also provide this technology to countries with underfunded infrastructure, as a repayment of Australia's climate debt for its historically high emissions. ...

A safe climate is not possible unless an informed and mobilised community fights for it. Australia's powerful fossil fuel industry has shown

it will not accept these measures. For years, it has backed the climate deniers in both major parties to prevent change."
Socialist Alliance - Climate Change Charter, January 16th 2010
We need to change the system, not the climate

"Climatologists, who study weather patterns over time, almost universally endorse the view that the earth is warming and that humans have contributed to climate change. There is less of a consensus among meteorologists, who predict short-term weather patterns.

Joe Bastardi, for example, a senior forecaster and meteorologist with AccuWeather, maintains that it is more likely that the planet is cooling, and he distrusts the data put forward by climate scientists as evidence for rising global temperatures.

Such skepticism appears to be widespread among TV forecasters, about half of whom have a degree in meteorology. A study released on Monday by researchers at George Mason University and the University of Texas at Austin found that only about half of the 571 television weathercasters surveyed believed that global warming was occurring and fewer than a third believed that climate change was 'caused mostly by human activities.'

More than a quarter of the weathercasters in the survey agreed with the statement 'Global warming is a scam,' the researchers found."
New York Times, March 29th 2010
Among Weathercasters, Doubt on Warming

"Well-publicized troubles have mounted for those forecasting global warming. First, there was last year's release of hacked e-mails from the United Kingdom's University of East Anglia, showing some climate scientists really dislike their critics (investigations are still ongoing). Then there was the recent discovery of a botched prediction that all Himalayan glaciers would disappear by 2035 in one of the Nobel-Prize-winning 2007 Intergovernmental Panel on Climate Change (IPCC) reports. ...

The recent controversies 'have really shaken the confidence of the public in the conduct of science,' according to atmospheric scientist Ralph Cicerone, head of the U.S. National Academy of Sciences. Cicerone was speaking at the American Association for the Advancement of Science meeting last month on a panel calling for more communication and release of data to rebuild lost trust for scientists. IPCC chiefs have made similar calls in the handling of their reports."

Dan Vergano,
USA Today, March 5th 2010
Some scientists misread poll data on global warming controversy

"The volcanic eruption in Iceland that disrupted air travel in Europe on Thursday was not a particularly powerful one, experts said, but they cautioned that its effects — both on travel and on the regional climate — might linger."

Henry Fountain,
The New York Times, April 15th 2010
Eruption Wasn't That Powerful, but Effects May Linger

"The eruption is now thought to have disrupted the Asian monsoon cycle, prompting famine in Egypt. Environmental historians have also pointed to the disruption caused to the economies of northern Europe, where food poverty was a major factor in the build-up to the French revolution of 1789.

Volcanologists at the Open University's department of earth sciences say the impact of the Laki eruptions had profound consequences.

Dr John Murray said: 'Volcanic eruptions can have significant effects on weather patterns for from two to four years, which in turn have social and economic consequences. We shouldn't discount their possible political impacts.'"

Greg Neale is founding editor of BBC History Magazine
The Guardian, April 15th 2010
How an Icelandic volcano helped spark the French Revolution

"If past is prelude, then the volcanic eruption in Iceland whose plume of ash has grounded almost 300 flights across Europe may not only affect air travel in the coming days, it may also have a lingering impact on Europe's weather. Experts are looking back to the aftereffects of a previous eruption—when the Laki volcano in Southern Iceland exploded more than 200 years ago. That explosion had catastrophic consequences for weather, agriculture and transport across the northern hemisphere – and helped trigger the French revolution."

Aline Reynolds,
Discover Magazine, April 16th 2010
Icelandic Volcanoes–Disrupting Weather & History Since 1783

"John P. Holdren, who then-President-elect Barack Obama nominated as director of the White House Office of Science and Technology Policy in December 2008, called just five months before his nomination for a global climate-change agreement that would allow wealth to be redistributed from countries in the global 'North' to countries in the 'South.'

On the July 3, 2008 edition of the program 'Democracy NOW!' Holdren told host Amy Goodman: 'It's important that we have a global agreement on how we are going to limit the emissions of carbon dioxide and other greenhouse gases going forward, and an agreement that will include the tropical forests, that will include ways to transfer some of the revenues from carbon taxes or carbon emission permits in the North to pay for reduced deforestation in the South.'"

Christopher Neefus,
CNS News, July 7th 2010
Obama Science Czar Called for Carbon Tax to Redistribute Wealth from Global 'North' to 'South'

"In Living in Denial, sociologist Kari Norgaard searches for answers to this question, drawing on interviews and ethnographic data from her study of "Bygdaby," the fictional name of an actual rural community in western Norway, during the unusually warm winter of 2000-2001.

Norgaard traces this denial through multiple levels, from emotions to cultural norms to political economy. Her report from Bygdaby, supplemented by comparisons throughout the book to the United States, tells a larger story behind our paralysis in the face of today's alarming predictions from climate scientists."
Kari Marie Norgaard
Amazon, March 11th 2011
Living in Denial: Climate Change, Emotions, and Everyday Life/MIT Press

"Nobody has ever offered a more succinct indictment of the global warming hoax than H. L. Mencken, who said: 'The whole aim of practical politics is to keep the populace alarmed (and hence clamorous to be led to safety) by menacing it with an endless series of hobgoblins, all of them imaginary.'"
F. Swemson,
American Thinker, July 16th 2011
The Global Warming Hoax, How Soon We Forget

"Across the nation, too, belief in man-made global warming, and passion about doing something to arrest climate change, is not what it was five years or so ago, when Al Gore's movie had buzz and Elizabeth Kolbert's book about climate change, 'Field Notes From a Catastrophe,' was a best seller. The number of Americans who believe the earth is warming dropped to 59 percent last year from 79 percent in 2006, according to polling by the Pew Research Group. When the British polling firm Ipsos Mori asked Americans this past summer to list their three most pressing environmental worries, 'global warming/climate change' garnered only 27 percent, behind even 'overpopulation.'"
Elisabeth Rosenthal,
The New York Times, October 15th 2011
Where Did Global Warming Go?

"Are you a global warming skeptic? There are plenty of good reasons why you might be. As many as 757 stations in the United States recorded net surface-temperature cooling over the past century. Many are

concentrated in the southeast, where some people attribute tornadoes and hurricanes to warming. The temperature-station quality is largely aweful."
Ruchard Muller,
The Wall Street Journal, October 21st 2011
The Case Against Global-Warming Skepticism

"While Americans are rightfully focused on the unemployment situation and the debt limit negotiations, we've pretty much forgotten about global warming as an issue ever since Obama failed to pass his Cap & Trade bill. As a result, we're becoming complacent once again about the huge threat we face from the progressives' attempts to control the world's energy industry based on the greatest scientific hoax in human history. In reality, however, nothing's changed, as Obama is still imposing his will on us through the EPA's regulation of CO2.

The truth is that CO2 is a beneficial trace gas that exists in such small quantities in our atmosphere, that the idea of it playing any significant role in determining our climate is simply silly. CO2 comprises less than half of 0.1% of our atmosphere, and only 4% of it comes from human activity. That's 16ppm, or 1 part in every 62,500 parts of our atmosphere. CO2 is plant food, and a key component in all life on earth. Plants need CO2 to grow and produce oxygen. They feed animals (including ourselves). Animals in turn consume oxygen and plant-based foods, and exhale CO2. Without CO2, nothing could be green! This brief video showing the effect on plants of increasing atmospheric CO2 is quite striking."
F. Swemson,
American Thinker, July 16th 2011
The Global Warming Hoax, How Soon We Forget

"Linda Sterio remembers the excitement when President Obama arrived at Solyndra last year and described how his administration's financial support for the plant was helping create hundreds of jobs. The company's prospects appeared unlimited as Solyndra executives described the backlog of orders for its solar panels.

Then came the August morning when Sterio heard a newscaster announce that more than a thousand Solyndra employees were out of work. Only recently did she learn that, within the Obama administration, the company's potential collapse had long been discussed.

'It's not about the people; it's politics,' said Sterio, who remains jobless and at risk of losing her home. 'We all feel betrayed.'

Since the failure of the company, Obama's entire $80 billion clean-technology program has begun to look like a political liability for an administration about to enter a bruising reelection campaign."
Joe Stephens and Carol D. Leonnig,
The Washington Post, December 25th 2011
Solyndra: Politics infused Obama energy programs

"The basic physics of climate change have been known for more than a century, but it is in recent decades that the fundamental science of global warming has solidified."
Spencer R.Weart,
Scientific American, August 17th 2012
The Discovery of Global Warming
Excerpt from Harvard University Press

"... Joe Kernen went off on people who believe in climate change. He made these comments during an interview with a former president and CEO at Shell Oil, who was pushing the government to embrace natural gas. Kernen told him it won't happen anytime soon, remarking that the government is currently too obsessed with climate change to deal with that issue. He used the occasion to go after the 'bonafide cult' of climate change 'enviro-socialists.'"
Josh Feldman,
Mediaite, June 6th 2013
CNBC's Joe Kernen Goes On Tirade Against 'Bonafide Cult' Of Climate Change: 'Enviro-Socialists'

"As the president demonstrated once again during his 'climate action plan' address in Georgetown, he is not someone ever to allow facts to stand in the way of ideology and green lobby cronyism. The familiar take-away line is that even more regulation is essential to bludgeon energy producers and consumers to abandon climate-ravaging fossil fuels in favor of heavily taxpayer-subsidized 'alternatives'.

Even his staunch allies in all things liberal at the New York Times appear to finally recognize that the feverish climate fervor behind these green grab gambits is overheated. They reported on June 6 that, 'The rise in the surface temperature of Earth has been markedly slower over the last 15 years than in the 20 years before that. And that lull in warming has occurred even as greenhouse gases have accumulated in the atmosphere at a record pace.' Reporter Justin Gillis went on to admit that the break in temperature increases 'highlights important gaps in our knowledge of the climate system,' whereby the lack of warming 'is a bit of a mystery to climate scientists.'"

Larry Bell,
Forbes, July 2nd 2013
Even The New York Times Has Chilled On Global Warming. Someone Please Tell Obama

"'It's a real problem ... it shows that there really is something that needs to be fixed in the climate models,' climate scientist John Christy, a professor at the University of Alabama in Huntsville, told FoxNews.com. ... Christy agrees that there has been some warming over time, but says man-made greenhouse gasses are not as big of a driver of climate change as many think -- and that many scientists are in denial about their mistakes. ... 'I think in one sense the climate establishment is embarrassed by this, and so they're trying to minimize the problem,' he said. 'The fundamental thing a climate model is supposed to predict is temperature. And yet it gets that wrong.'"

Maxim Loft,
Fox News, September 2nd 2013
Climate models wildly overestimated global warming, study finds

"Many of the overestimations also made their way into the popular press. In 1989, the Associate Press reported: 'Using computer models, researchers concluded that global warming would raise average annual temperatures nationwide 2 degrees by 2010.' ... But according to NASA, global temperature has increased by less than half that -- about 0.7 degrees Fahrenheit -- from 1989 to 2010."
Maxim Loft,
Fox News, September 2nd 2013
Climate models wildly overestimated global warming, study finds

"Al Gore and his shrinking band of far left lunatics have now been proven wrong merely by comparing their claims to what has actually happened. The truth is crystal clear and right in front of us."
Dr. Kevin Collins,
Western Journalism, September 19th 2013
Global warming predictions proven wrong 97.4% of the time

"As for letters on climate change, we do get plenty from those who deny global warming. And to say they 'deny' it might be an understatement: Many say climate change is a hoax, a scheme by liberals to curtail personal freedom."
Paul Thornton,
Los Angeles Times, October 8th 2013
On Letters from Climate Change Deniers

"The Los Angeles Times has stirred a dust-up over global warming with a newly announced policy barring letters to the editor that deny the existence of man-made climate change."
Valerie Richardson,
The Washington Times, October 10th 2013
L.A. Times nixes letters from climate change deniers

"For many who advocate costly responses to 'irrefutable evidence' that the world's climate faces catastrophe, global warming has become a

substitute religion. Increasingly offensive language is used: the most egregious example being the term 'denier'. We all know the particular meaning that word has acquired in contemporary parlance. It has been employed in this debate with malice aforethought."
John Howard,
U.K. Telegraph, November 9th 2013
Climate change is an uncertain science
You don't have to be a 'denier' to question many of the West's short-sighted policies on renewable energy

"Dr. Don Easterbrook – a climate scientist and glacier expert from Washington State who correctly predicted back in 2000 that the Earth was entering a cooling phase – says to expect colder temperatures for at least the next two decades. Easterbrook's predictions were 'right on the money' seven years before Al Gore and the United Nation's Intergovernmental Panel on Climate Change (IPCC) shared the 2007 Nobel Peace Prize for warning that the Earth was facing catastrophic warming caused by rising levels of carbon dioxide, which Gore called a 'planetary emergency.' 'When we check their projections against what actually happened in that time interval, they're not even close. They're off by a full degree in one decade, which is huge. That's more than the entire amount of warming we've had in the past century. So their models have failed just miserably, nowhere near close. And maybe it's luck, who knows, but mine have been right on the button,' Easterbrook told CNSNews.com."
Barbara Hollingsworth
CNS News, January 28th 2014
Climate Scientist Who Got It Right Predicts 20 More Years of Global Cooling

"BRITAIN could freeze in YEARS of super-cold winters and miserable summers if the Bardarbunga volcano erupts, experts have warned. ...

The 1980 eruption of Mount St Helens in Skamania County, Washington, United States, led to global temperatures dropping by 0.1C."
Nathan Rao,
U.K. Express, August 24th 2014

Icelandic volcano could trigger Britain's coldest winter EVER this year

"The sulfur dioxide (SO2) emitted from the Holuhraun eruption has reached up to 60,000 tons per day and averaged close to 20,000 tons since it began. For comparison, all the SO2 pollution in Europe, from industries, energy production, traffic and house heating, etc., amounts to 14,000 tons per day."
Pall Stefanson,
Iceland Review Online , September 25th 2014
Holuhraun Emitting More SO2 Pollution than All of Europe

"Our planet has about 1,000 volcanoes on land, such as Holuhraun and Bardarbunga, but most of our volcanoes are under the sea. 'Some 85 per cent of volcanoes are unseen and unmeasured yet these heat the oceans and add monstrous amounts of CO2 to the oceans,' notes Dr. Plimer. 'Why have these been ignored?,' he asks.

In a video lecture at the Institute of Geology and Geophysics at the University of Adelaide (which can be viewed here), Professor Plimer notes that the more than 10,000 earthquakes that occur each year release massive amounts of CO2 that has been sequestered in the various mineral formations. CO2 is but one of many variables that affect the climate, and its effect is very slight, in comparison to many other factors. Moreover, man's contribution of CO2, SO2, and other greenhouse gases is relatively minor, when compared to the contributions from natural sources."
William F. Jasper,
The New American, September 26th 2014
Iceland's Volcanic Pollution Dwarfs All of Europes Human Emissions

"Most Americans think global warming poses a critical threat — in the future, for other people. A majority said it will be a very serious problem for the future of the world, but fewer described it as very serious for the United States."
The New York Times, January 29th 2015
Global Warming: What Should Be Done?

"The severity of the 2006 to 2010 drought, and more importantly the failure of Bashar al-Assad's regime to prepare, or respond to it effectively, exacerbated other tensions, from unemployment to corruption and inequality, which erupted in the wake of the Arab spring revolutions, the scientists say.

'We're not arguing that the drought, or even human-induced climate change, caused the uprising,' said Colin Kelley at the University of California in Santa Barbara. 'What we are saying is that the long term trend, of less rainfall and warmer temperatures in the region, was a contributing factor, because it made the drought so much more severe.'"
Ian Sample,
The Guardian, March 2nd 2015
Global warming contributed to Syria's 2011 uprising, scientists claim

"Despite being bombarded with warnings from environmentalists and politicians, Americans still aren't very worried about global warming. A new Gallup poll shows that Americans' concern about warming has fallen to the same level it was in 1989. In fact, global warming ranked at the bottom of a list of Americans' environmental concerns, with only 32 percent saying they worried about it a 'great deal.'"
Michael Bastasch,
Daily Caller, March 25th 2015
Poll: America's Fear Of Global Warming Drops To 1980s Levels

"Oddly enough, the solutions scientists pushed to fight global cooling and warming are the same: ban fossil fuels and use low-carbon energy."
Michael Bastasch
Daily Caller, April 3rd 2015
Flashback 1971: Scientists Predict Burning Coal Will Cause The Next Ice Age

"In the May 2000 issue of Reason Magazine, award-winning science correspondent Ronald Bailey wrote an excellent article titled 'Earth Day,

Then and Now' to provide some historical perspective on the 30th anniversary of Earth Day. ...

How accurate were the predictions made around the time of the first Earth Day in 1970? The answer: 'The prophets of doom were not simply wrong, but spectacularly wrong,' according to Bailey.

1. Harvard biologist George Wald estimated that 'civilization will end within 15 or 30 years unless immediate action is taken against problems facing mankind.' ...

7. 'It is already too late to avoid mass starvation,' declared Denis Hayes, the chief organizer for Earth Day, in the Spring 1970 issue of The Living Wilderness. ...

9. In January 1970, Life reported, 'Scientists have solid experimental and theoretical evidence to support...the following predictions: In a decade, urban dwellers will have to wear gas masks to survive air pollution...by 1985 air pollution will have reduced the amount of sunlight reaching earth by one half...' ...

18. Kenneth Watt warned about a pending Ice Age in a speech. 'The world has been chilling sharply for about twenty years,' he declared. 'If present trends continue, the world will be about four degrees colder for the global mean temperature in 1990, but eleven degrees colder in the year 2000. This is about twice what it would take to put us into an ice age.'"

Mark J. Perry,
AEIdeas, April 21st 2015
18 spectacularly wrong apocalyptic predictions made around the time of the first Earth Day in 1970, expect more this year

"Our world is fundamentally changing. Fourteen of the 15 warmest years on record have occurred since 2000. Last year was the warmest of all.

And with added heat comes an altered environment. It's difficult to determine whether one specific storm or drought is solely caused by climate change, but the growing intensity of storms and changing weather patterns should be a clear signal. Here in the United States, California is enduring the fourth year of its worst drought in recorded history.

This is just the beginning. If we don't make significant changes — quickly — scientists say we can expect sea levels to continue rising to dangerous levels, more intense and frequent extreme weather events, severe disruptions to food supplies and prolonged resource shortages."
John Kerry,
USA Today, April 22nd 2015
John Kerry: On Earth Day, time running out for climate change

"Finally, think about this question, posed by Ronald Bailey in 2000: What will Earth look like when Earth Day 60 rolls around in 2030? Bailey predicts a much cleaner, and much richer future world, with less hunger and malnutrition, less poverty, and longer life expectancy, and with lower mineral and metal prices. But he makes one final prediction about Earth Day 2030: 'There will be a disproportionately influential group of doomsters predicting that the future–and the present–never looked so bleak.' In other words, the hype, hysteria and spectacularly wrong apocalyptic predictions will continue, promoted by the 'environmental grievance hustlers.'"
Mark J. Perry,
AEIdeas, April 21st 2015
18 spectacularly wrong apocalyptic predictions made around the time of the first Earth Day in 1970, expect more this year

"'The science in 1981 on this subject was in the very, very early days and there was considerable division of opinion,' Richard Keil, an Exxon spokesman, said. 'There was nobody you could have gone to in 1981 or 1984 who would have said whether it was real or not. Nobody could provide a definitive answer.'

He rejected the idea that Exxon had funded groups promoting climate denial. 'I am here to talk to you about the present' he said."
The Guardian, July 8th 2015

Exxon knew of climate change in 1981, email says – but it funded deniers for 27 more years

"The president believes that 'there is no global warming, that this is a fraud to restrain the industrial development of several countries including Russia,' says Stanislav Belkovsky, a political analyst and critic of Putin. 'That is why this subject is not topical for the majority of the Russian mass media and society in general.' ...

Putin's scepticism dates from the early 2000s, when his staff 'did very, very extensive work trying to understand all sides of the climate debate', said Andrey Illarionov, Putin's senior economic adviser at the time and now a senior fellow at the Cato Institute in Washington.

'We found that, while climate change does exist, it is cyclical, and the anthropogenic role is very limited,' he said. 'It became clear that the climate is a complicated system and that, so far, the evidence presented for the need to 'fight' global warming was rather unfounded.'"
Reuters – World News, October 29th 2015
Russian media take climate cue from skeptical Putin

"'Representative democracy' has failed; the private sector is 'inept'; and only bigger government – led by China and the US – has the power to save the world from climate change.

So says Bill Gates in a dogmatic but somewhat confused interview with The Atlantic in which he simultaneously pours scorn on green tech solutions but insists that more of them are needed – on a scale bigger than the Manhattan Project – if we are to deal successfully with a problem whose nature he admits may well have been exaggerated by environmentalists."
James Delingpole,
Breitbart, November 2nd 2015
Bill Gates: Only Socialism Can Save Us From Climate Change

"It was a time of yuppies, flash cars, shoulder pads and big hair, but it appears the 1980s was also a key turning point for the world's climate, research has suggested. Scientists have discovered there was a huge shift in the environment that swept across the globe affecting ecosystems from the depths of the oceans to the upper atmosphere. They said an abrupt spurt of global warming, fuelled by human activity and a volcanic eruption in Mexico, is believed to have triggered these changes between 1984 and 1988."

Richard Gray,

Daily Mail UK, November 24th 2015

How a volcanic eruption in the 1980s triggered a 'spurt' of global warming: Event caused a shift in Earth's climate that may have killed off several species of animal

"Planet Earth experienced a global climate shift in the late 1980s on an unprecedented scale, fuelled by anthropogenic warming and a volcanic eruption, according to new research published this week.

Scientists say that a major step change, or 'regime shift', in Earth's biophysical systems, from the upper atmosphere to the depths of the ocean and from the Arctic to Antarctica, was centered around 1987, and was sparked by the El Chichón volcanic eruption in Mexico five years earlier."

Science Daily, November 24th 2015

Climate study finds evidence of global shift in the 1980s

Anthropogenic warming, volcanic eruption sparked biggest change in 1,000 years

"In January, 2006 — when promoting his Oscar-winning (yes, Oscar-winning) documentary, An Inconvenient Truth — Gore declared that unless we took 'drastic measures' to reduce greenhouse gasses, the world would reach a 'point of no return' in a mere ten years. He called it a 'true planetary emergency.' Well, the ten years passed today, we're still here, and the climate activists have postponed the apocalypse. Again.

Gore's prediction fits right in with the rest of his comrades in the wild-

eyed environmentalist movement. There's a veritable online cottage industry cataloguing hysterical, failed predictions of environmentalist catastrophe. Over at the American Enterprise Institute, Mark Perry keeps his list of '18 spectacularly wrong apocalyptic predictions' made around the original Earth Day in 1970. Robert Tracinski at The Federalist has a nice list of 'Seven big failed environmentalist predictions.' The Daily Caller's '25 years of predicting the global warming 'tipping point'' makes for amusing reading, including one declaration that we had mere 'hours to act' to 'avert a slow-motion tsunami.'"
David French,
National Review, January 27th 2016
Apocalypse Delayed

"The debate between researchers and doubters reached a crescendo last summer, when scientists at the National Oceanic and Atmospheric Administration updated their temperature records and concluded that global warming has not slowed down in the 2000s (ClimateWire, June 5, 2015).

Now, a group of prominent climate scientists are challenging NOAA's conclusion in a commentary published this week in Nature Climate Change.

'The interpretation [the NOAA group] made was not valid,' said John Fyfe, a climate scientist at Environment and Climate Change Canada and lead author of the commentary. 'The slowdown is there, even in this new updated data set.'"
Gayathri Vaidyanathan,
Scientific American, February 25th 2016
Did Global Warming Slow Down in the 2000s, or Not?

"Carbon dioxide (CO_2) is a greenhouse gas and is the primary gas blamed for climate change. While sulfur dioxide released in contemporary volcanic eruptions has occasionally caused detectable global cooling of the lower atmosphere, the carbon dioxide released in contemporary volcanic

eruptions has never caused detectable global warming of the atmosphere."
U.S. Geological Survey, February 26th 2016
Volcanoes can affect the Earth's climate

"Volcanoes can impact climate change. During major explosive eruptions huge amounts of volcanic gas, aerosol droplets, and ash are injected into the stratosphere. Injected ash falls rapidly from the stratosphere -- most of it is removed within several days to weeks -- and has little impact on climate change. But volcanic gases like sulfur dioxide can cause global cooling, while volcanic carbon dioxide, a greenhouse gas, has the potential to promote global warming."
U.S. Geological Survey, February 26th 2016
Volcanoes can affect the Earth's climate.

"On the advent of socialism a considerable reduction in CO2 emissions would be obtained by ending the enormous wasteful economic activity and production inherent in capitalism. Principal contributors to this waste are the maintenance and preparedness of the armed forces and occupations and products handling money. The enormous sums of money spent on the military represent the use of land, facilities, machinery, materials and human resources to keep it in a constant state of preparedness for and participation in war. The Socialist Party's own publication From Capitalism to Socialism provides a list of dozens of occupations and products for the financial system as diverse as accountancy, advertising, banking, income tax officers, VAT inspectors, stock brokers, stock exchanges, banknotes, armoured cars, gas and electricity meters, postage stamps, tickets and TV licences. Socialism would eliminate the need for these industries and thereby rapidly introduce a large cut in power generation and consequent CO2 emissions to the atmosphere. ...

To conclude, in the first period of socialism clearing up this mess left by capitalism would be a priority project. However it is certainly the case that despite the reduction in CO2 emissions that would be brought about by socialism, the introduction of the wedge strategy would still be a

necessary and formidable challenge, but surely one that would be grasped wholeheartedly in a sane, socialist world."
Social Standard on WorldSocialism.org, May 2016
Socialism and Climate Change

"America will never be destroyed from the outside. If we falter and lose our freedoms, it will be because we destroyed ourselves."
Abraham Lincoln

"Alberta's new climate plan, by the government's own admission, will not lead to significant greenhouse gas reductions for many years, if it ever does. The tax is far too low to have significant impacts on consumer behaviour, which is further proof that it is not 'market-based carbon pricing,' it's a funding mechanism for bureaucratic expansion of failed efficiency programs, a mechanism for redistributing the wealth of Albertans, and a green fig leaf for an Alberta government that wants to look as if it's proactively promoting 'social license' for the continued development of Alberta's oilsands. Alberta's new carbon tax should be called what it is: Alberta's new Provincial Sales Tax."
Kenneth P. Green,
Fraser Institute, June 6th 2016
Alberta carbon tax will fund bureaucratic expansion, redistribute wealth of Albertans

*"Scientists don't fully understand why Antarctic sea ice is growing —
suggested explanations have posited more glacial melt dumping cold fresh water into the surrounding seas, or the way the Antarctic ozone hole has changed the circulation of winds around the continent. ...*

The new study confirms that the ice floating around Antarctica has been expanding — indeed, the expansion has accelerated since around the turn of the century. ...

But the new study finds that in the small minority of climate change simulations that do happen to correctly capture these natural changes in the Pacific, and the global warming 'slowdown' to boot, there is also growth in Antarctic sea ice. These are the models, it appears, that happened to get the role of natural variability in the Pacific right — or more specifically, to get the timing right for a phase shift in this ocean."
Chris Mooney,
Washington Post, July 5th 2016
This new Antarctica study is bad news for climate change doubters

"'Democrats believe that climate change is too important to wait for climate deniers in Congress to start listening to science,' Houser said. ...

The Democratic-led House passed a 'BTU tax' — for British thermal unit, a measure of energy — in 1993, costing the party its majority the next year. After Democrats retook the House, they passed a cap-and-trade bill in 2009, and again lost the majority in 2010.

Conservative anti-tax advocate Grover Norquist, who leads Americans for Tax Reform, said the Democrats probably lost the 2016 election already by including a carbon tax in their platform."
Timothy Cama,
The Hill, July 31st 2016
Clinton walks fine line on carbon tax

"Climate disruption is inextricably linked to economic inequality. Serious climate solutions must be, too. ...

A carbon tax could help transfer wealth from people at the carbon-intensive top to less polluting middle and lower-income households, and ensure the costs of addressing climate change are distributed equitably."
The Institute for Policy Studies, August 10th 2016
To Stop Climate Change, Don't Just Cut Carbon. Redistribute Wealth.

"Socialist Action's vice-presidential candidate Karen Schraufnagel has filed extensive reports to this paper about the confrontations last month in remote Standing Rock, North Dakota. Efforts aiming to halt construction of a pipeline from the Bakken fracking fields to refineries in Illinois drew world attention to several important issues concerning Indigenous People's rights, environmental and climate justice, and peaceful protests under violent attack."
BILL ONASCH
Socialist Action.org, October 10th 2016
Taking Sides at Standing Rock

"None of the revenues from pricing carbon would be used to increase investment in clean energy, build climate resiliency or create green jobs — particularly in communities on the frontlines of climate change. ...

Supporters say we can't let the perfect be the enemy of the good in the face of the urgent need to address climate disruption. But a measure like I-732 that ignores the urgency of environmental justice and economic fairness actually makes it harder to advance our long-term goals of a clean environment and green economy. Because of this, the measure lacks support from labor, environmental justice groups and communities of color, all of whom are crucial leaders in the fight against climate change and for a more equitable economy."
Bill Corcoran and Byron Gudiel,
PBS, November 8th 2016
Washington's carbon tax doesn't address environmental justice

"'North Korea just stated that it is in the final stages of developing a nuclear weapon capable of reaching parts of the U.S. It won't happen!' Trump tweeted."
Nicole Gaouette and Barbara Starr,
CNN, January 3rd 2017
Facing growing North Korea nuke threat, Trump vows:'It won't happen!'

Interesting comparison between the last quote and the next...

"North Korea's Ministry of Foreign Affairs called global warming 'one of the gravest challenges humankind is facing today.'"
Fox News, June 6th 2017
North Korea slams Trump for pulling out of Paris Climate Agreement

"Still, there remain big pockets of climate confusion -- perhaps denial -- across the country, especially when it comes to climate science."
John D. Sutter,
CNN, February 28th 2017
Common ground with climate skeptics

"Environmental Protection Agency chief Scott Pruitt on Thursday doubled down on climate change denial, saying he doesn't believe carbon dioxide is to blame for global warming.

'I think that measuring with precision human activity on the climate is something very challenging to do and there's tremendous disagreement about the degree of impact, so no, I would not agree that it's a primary contributor to the global warming that we see,' he told CNBC. 'We need to continue the debate and continue the review and the analysis.'"
Alana Horowitz Satlin,
Huffington Post, March 9th 2017
EPA Chief Scott Pruitt Disagrees With Science On Another Major Climate Change Issue

"Conservatives have long been the voice for limited government, lower taxes, free markets, and individual liberty. But recently, a small but persistent group of Republicans are trying to persuade conservatives to abandon these principles and embrace a national energy tax.

The idea of taxing carbon isn't new. Bill Clinton and Al Gore tried to pass a BTU tax in 1993. Its defeat helped usher in a Republican-controlled Congress for the first time in nearly 40 years. In 2009, Barack Obama proposed a cap-and-trade plan that was rejected by a Democratic-controlled Congress. The Democrats ended up losing the House in 2010.

More recently, Hillary Clinton's campaign policy team was reportedly considering a carbon tax. ...

A carbon tax would punish users of natural gas, oil, and coal, which make up 80 percent of the energy we consume. This means that all American families would face higher electricity bills and gasoline prices. In fact, it's estimated that the Council's carbon tax would hike gasoline prices by 36 cents per gallon. While everybody will pay more, these hikes would have a disproportionate impact on poor and middle-class families, who spend a higher percentage of their income on energy. It also means Americans would pay more for goods and services across the board."
Thomas Pyle,
National Review, March 23rd 2017
There's Nothing Conservative about a Carbon Tax

"Republicans and three scientists accused mainstream climate scientist, Michael Mann of Pennsylvania State University, and major international science panels of trying to quiet researchers who disagree about the magnitude of global warming.

Mann and Republican Rep. Dana Rohrabacher of California both compared the other side's behavior to the former Soviet Union under Josef Stalin. Mann first raised the Stalin analogy, then Rohrabacher used the comparison four times after that to talk about Mann and other mainstream climate scientists.

'For scientists to call names to beat someone into submission, that's a Stalinist tactic,' Rohrabacher said."
Seth Borenstein,
Yahoo News - Associated Press, March 29th 2017
US hearing on climate science focuses on name calling

"Former Georgia Tech climate scientist Judith Curry, who often clashes with mainstream science, said she was the victim of 'gutter tactics' by 'scientists who demonize their opponents.'

She pointed to the way she and Mann have clashed, saying the Penn State professor wrongly called her a climate denier, when she acknowledges that the world is warming and humans play a role. She disagrees with mainstream climate science over implications of global change, the size of the warming, how much is human-caused and its certainty.

At first Mann said he didn't call Curry a denier. But in his written not oral testimony he called Curry 'a climate science denier.' Mann said there's a difference between denying climate change and 'denying established science' on how much humans cause climate change, which he said Curry did."

Seth Borenstein,
Yahoo News - Associated Press, March 29th 2017
US hearing on climate science focuses on name calling

"Record-breaking weather events, especially heat waves but also downpours and droughts, can be linked to man-made global warming, a new study says. ...

'Our results suggest that the world isn't quite at the point where every record hot event has a detectable human fingerprint, but we are getting close,' said study lead author Noah Diffenbaugh, a climate scientist at Stanford University. ...

It's the first research to look specifically at the link between record weather events of the past several decades and climate change. Diffenbaugh and his team found that in over 80% of the heat records — which included both record hot days and months - there was a clear-cut signal of global warming."

Doyle Rice
USA Today, April 24th 2017
Global warming blamed for record-breaking weather worldwide, scientists say

"The activist does her best to present fact-based evidence that the climate movement deserves the man's support, but she has to allow that while there is consensus on the basic mechanisms of climate change, there have been some contradictory studies. That's how science works. It's why peer reviews are important and why researchers keep inquiring, testing theses, and adding to the canon. Scientists who study cancer and its causes also don't always agree, but that's not a reason to call off cancer research funding.

The merits of a cost-benefit analysis will only go so far in arguing for action to address climate action, and in the end, the authors explain, we need to make some qualitative judgments. Acknowledge gaps in knowledge—and remember that it doesn't defeat your purpose."
Mary Catherine O'Connor,
Outside Online, Apr 26th 2017
How to Reason with the Climate Change Denier in Your Life

"There's a lesson here. We live in a world in which data convey authority. But authority has a way of descending to certitude, and certitude begets hubris. From Robert McNamara to Lehman Brothers to Stronger Together, cautionary tales abound.

We ought to know this by now, but we don't. Instead, we respond to the inherent uncertainties of data by adding more data without revisiting our assumptions, creating an impression of certainty that can be lulling, misleading and often dangerous.

Let me put it another way. Claiming total certainty about the science traduces the spirit of science and creates openings for doubt whenever a climate claim proves wrong. Demanding abrupt and expensive changes in public policy raises fair questions about ideological intentions. Censoriously asserting one's moral superiority and treating skeptics as imbeciles and deplorables wins few converts."
Bret Stephens,
The New York Times, April 28th 2017
Climate of Complete Certainty

"In his first column for the Times, Bret Stephens said advocates for climate policy can take a lesson from Hillary Clinton's failed presidential campaign and her reliance on data to predict the election.

'We live in a world in which data convey authority. But authority has a way of descending to certitude, and certitude begets hubris,' Stephens wrote. 'Claiming total certainty about the science traduces the spirit of science and creates openings for doubt whenever a climate claim proves wrong.'"
Jackie Wattles and Dylan Byers
CNN Money, May 1st 2017
New York Times faces wave of criticism over climate column

"Climate change is real. There will always be uncertainty in understanding a system as complex as the world's climate. ...

Climate-warming trends over the past century are extremely likely due to human activities."
Earth Science Communications Team at NASA, June 1st 2017
Scientific consensus: Earth's climate is warming

"Coulter said all these people may talk a big game global warming, but their actions go to show that they either don't believe in it or they simply don't care.

'If you tell us that CO2 emissions are destroying the world and you're flying a private jet, it's obviously not about CO2 emissions for you,' Tucker said. 'What is it really about?'"
Fox Insider, June 2nd 2017
Coulter Slams Hillary, Obama & Others for Their Hypocrisy on Climate Change

"Like every other major problem confronting mankind, climate change is fundamentally a class question. It is the working class that will suffer the brunt of the impact of global warming. It is the working class that is

objectively and increasingly defining itself as an international class. It is the working class whose social interests lie in the overthrow of capitalism, the abolition of private ownership of the means of production, and the establishment of an economic system based on the satisfaction of human need, including a safe and heathy environment. ...

The dangers posed by global warming can be addressed only through a political struggle by the international working class against the anarchic and backward capitalist mode of production. Only in this way can the world's economy be rationally and scientifically reorganized and an environmental catastrophe averted. In short, the solution to climate change is socialism."
Bryan Dyne,
World Socialist Website, June 3rd 2017
International Committee of the Fourth International
Trump's withdrawal from the Paris agreement: The socialist solution to climate change

"The Party seeks power entirely for its own sake. We are not interested in the good of others; we are interested solely in power, pure power."
George Orwell, 1984

"It's been more than a decade since Al Gore predicted 'a true planetary crisis' due to global warming. With that failing to happen, however, Fox News's Chris Wallace figured the former vice president had some explaining to do.

In an interview Sunday, Wallace confronted Gore over claims he made in his 2006 documentary 'An Inconvenient Truth,' including that unless 'took drastic measures [were taken] the world would reach a point of no return within 10 years.'

'Weren't you wrong?' Wallace asked.

'No,' Gore replied. 'Well we have seen a decline in emissions on a global basis. For the first time they've stabilized and started to decline. So some of the responses for the last 10 years have helped, but unfortunately and regrettably a lot of serious damage has been done.' ....

Gore also claimed in his 2006 film that by 2016, Mount Kilimanjaro in Africa would be snow-free and that weather would worsen, with stronger, more frequent hurricanes. Yet these, among other predictions, never happened."
Townhall, June 5th 2017
Al Gore Grilled Over Climate Predictions That Never Happened

"Power is in tearing human minds to pieces and putting them together again in new shapes of your own choosing."
George Orwell, 1984

"JESSE WATTERS (CO-HOST): 'People are dying from terrorism. No one is dying from climate change.'"
Media Matters, June 5th 2017
Fox News' Jesse Watters: "No one is dying from climate change"

"'Global warming' is a myth — so say 80 graphs from 58 peer-reviewed scientific papers published in 2017. In other words, the so-called 'Consensus' on global warming is a massive lie. And Donald Trump was quite right to quit the Paris agreement which pretended that the massive lie was true."
Breitbart, June 6th 2017
DELINGPOLE: 'Global Warming' Is a Myth, Say 58 Scientific Papers in 2017

"'In order to fulfill my solemn duty to protect America and its citizens, the United States will withdraw from the Paris climate accord,' Mr. Trump said, and instead, he said that the U.S. will 'begin negotiations to reenter

either the Paris accord or really an entirely new transaction' on terms that he said are fair to U.S. businesses, workers, and taxpayers."
Rebecca Shabad,
CBS News, June 2nd 2017
Trump Pulls out of Climate Deal

"CNBC's Joe Kernen got into some Twitter back-and-forths on the subject of climate change this week, even saying at one point that it could end up being "a socialist, global wealth redistribution scheme."
Josh Feldman,
Mediaite, June 6th 2017
CNBC's Joe Kernen: At Worst Climate Change Is 'A Socialist, Global Wealth Redistribution Scheme'

"At the root of our socialism is a profound commitment to democracy, as means and end. As we are unlikely to see an immediate end to capitalism tomorrow, DSA fights for reforms today that will weaken the power of corporations and increase the power of working people. ...

We are activists committed to democracy as not simply one of our political values but our means of restructuring society. ...

We are socialists because we reject an international economic order sustained by private profit, alienated labor, race and gender discrimination, environmental destruction, and brutality and violence in defense of the status quo."
Democratic Socialists of America, 2017
www.dsausa.org

"Power is not a means; it is an end. One does not establish a dictatorship in order to safeguard a revolution; one makes the revolution in order to establish the dictatorship. The object of persecution is persecution. The object of torture is torture. The object of power is power."
George Orwell, 1984

"Some of the fastest progress on clean energy is occurring in states led by Republican governors and legislators, and states carried by Donald J. Trump in the presidential election. …

'I think the answer is that we don't need these silly wars. Let's not even try to agree on climate change,' said Hal Harvey, chief executive of Energy Innovation, a think tank in San Francisco. 'Let's just get the job done.'"
Justin Gillis and Nadja Popovich,
The New York Times, June 6th 2017
In Trump Country, Renewable Energy Is Thriving

"A global warming research study in Canada has been cancelled because of 'unprecedented' thick summer ice.

Naturally, the scientist in charge has blamed it on 'climate change.' …

There was the Ship of Fools expedition in which an Australian climate researcher called Chris Turkey had to call an expedition to the melting Antarctic after his ship got stuck in the ice.

Most recently there was Ship of Fools II, in which a global warming research voyage by David Hempleman Adams had to be curtailed because of unexpected ice."
James Delingpole,
Breitbart, June 13th 2017
DELINGPOLE: Ship of Fools III – Global Warming Study Cancelled
Because of 'Unprecedented' Ice

BONUS – Gory Details

"Can Big Brother be green? Absolutely. If carbon dioxide were the planetary poison that global warming alarmists claim, then every aspect of our lives would be fair game for government control: the homes we build, the cars we drive, the light bulbs we use. Even the number of children we have—because lets face it; any reduction in CO_2 that we achieve will be more than offset by the households our kids will create when they grow up.

There are already proposals in Congress and federal agencies to vastly increase taxes and regulations in order to address the so-called global warming crisis. But as a growing number of scientists are openly declaring, there is no crisis. ...

Irony aficionados—In its original 1984 ad, Apple Computer warned of a totalitarian threat in computing. Today Al Gore sits on Apples Board of Directors. The company that warned of 1984 25 years ago now has, as one of its directors, the man most likely to lead us into a new 1984."
Prison Planet Online, May 15th 2009
Al Gore 1984

"Apple® today announced that Albert Gore Jr., the former Vice President of the United States, has joined the Company's Board of Directors. Mr. Gore was elected at Apple's board meeting today.

'Al brings an incredible wealth of knowledge and wisdom to Apple from having helped run the largest organization in the world—the United States government—as a Congressman, Senator and our 45th Vice President. Al is also an avid Mac user and does his own video editing in Final Cut Pro,' said Steve Jobs, Apple's CEO. 'Al is going to be a terrific Director and we're excited and honored that he has chosen Apple as his first private sector board to serve on.'"
Apple Newsroom, March 19th 2003
Former Vice President Al Gore Joins Apple's Board of Directors

"Two weeks after pulling down a reported $100 million from the sale of Current TV to the Qatar-based news network, the man so many love to hate is at it again, this time by exercising options entitling him to buy 59,000 Apple (Nasdaq: AAPL) shares for the excellent price of $7.475 each.

So Gore has just spent a little over $440,000 to acquire stock worth nearly $30 million as of Thursday's close. And what did he do for all that money? Attended a board meeting a few times a year for the last decade. Had Steve Jobs' back when it counted. That's all. Not bad for someone who didn't invent the Internet."
Forbes, January 18th 2013
Don't Hate On Al Gore For His $30 Million Apple Score

"A new study by a 'group of mean reactionary science purists' at the Journal Nature Climate Change took a look at 117 of the Left's wild claims. Its results are devastating for Gore and the Left. They show that 114, or 97.4%, of them were wrong. They weren't just a little 'ooops, I hit the wrong number on my computer – wrong but wrong by a you – don't – know what – in- the – world – you're –are talking – about' wrong. The average 'mistake' was an overstatement of global warming by double actual reality."
Dr. Kevin Collins
Western Journalism, September 19th 2013
Global warming predictions proven wrong 97.4% of the time

"How inconvenient is this news? Former vice president Al Gore and his wife, Tipper, bought a gated $8,875,000 ocean-view villa in Montecito, Calif., where Oprah Winfrey also owns a mansion, the Los Angeles Times reports."
Wendy Koch,
USA Today, May 18th 2010
How green is Al Gore's $9 million Montecito oceanfront villa?

CHAPTER 8

Vitrioloc Climate in Academic Hothouse

"Green politics have taken the place of failed socialism..." - Ian Plimer

The article below was written by Ian Pilmer, an Australian *professor of earth sciences,* and published in a magazine called *The Australian.* Personally I feel this is one of the most powerful articles I read on my journey to discover the truth about climate change and the policy behind the global climate agenda.

"IT is well known that many university staff list to port and try to engineer a brave new world. The cash cow climate institutes now seem to be drowning in their own self-importance.

In a wonderful gesture of public spiritedness, seven academics who include three lead authors of the Intergovernmental Panel on Climate Change and a former director of the World Climate Research Program wrote to Australian power generating companies on April 29 instructing them to cease and desist creating electricity from coal.

In their final paragraph, they state with breathtaking arrogance:

'The unfortunate reality is that genuine action on climate change will require the existing coal-fired power stations to cease operating in the near future.'

'We feel it is vital that you understand this and we are happy to work with you and with governments to begin planning for this transition immediately.'

'The warming of the atmosphere, driven by human-induced emissions of greenhouse gases, is already causing unacceptable damage and suffering around the world.'

No evidence is provided for this statement and no signatory to this letter has published anything to support this claim.

These university staff are unctuously understanding about the plight of those who face employment extinction in the smokestack towns of Australia.

They write: 'We understand that this will require significant social and economic transition that will need to be managed carefully to care for coal sector workers and coal-dependent communities.' This love for fellow workers brings tears to the eyes.

The electricity generating companies should reply by cutting off the power to academics' homes and host institutions, forcing our ideologues to lead by example.

Some 80 per cent of Australia's electricity derives from coal, large volumes of cheap electricity underpin employment and our self-appointed concerned citizens offer no suggestion for alternative unsubsidized base-load power sources to employ Australians.

The Emissions Trading Scheme legislation poises Australia to make the biggest economic decision in its history, yet there has been no scientific due diligence.

There has never been a climate change debate in Australia. Only dogma. To demonise element number six in the periodic table is amusing. Why not promethium? Carbon dioxide is an odourless, colourless, harmless natural gas. It is plant food. Without carbon, there would be no life on Earth.

The original source of atmospheric CO_2 is volcanoes. The Earth's early atmosphere had a thousand times the CO_2 of today's atmosphere. This CO_2 was recycled through rocks, life and the oceans.

Through time, this CO2 has been sequestered into plants, coal, petroleum, minerals and carbonate rocks, resulting in a decrease in atmosphericCO2.

The atmosphere now contains 800billion tonnes of carbon as CO2. Soils and plants contain 2000 billion tonnes, oceans 39,000 billion tonnes and limestone 65,000,000 billion tonnes. The atmosphere contains only 0.001 per cent of the total carbon in the top few kilometres of the Earth.

Deeper in Earth, there are huge volumes of CO2 yet to be leaked into the atmosphere. So depleted is the atmosphere in CO2, that horticulturalists pump warm CO2 into glasshouses to accelerate plant growth.

The first 50 parts per million of CO2 operates as a powerful greenhouse gas. After that, CO2 has done its job, which is why there has been no runaway greenhouse in the past when CO2 was far higher.

During previous times of high CO2, there were climate cycles driven by galactic forces, the sun, Earth's orbit, tides and random events such as volcanoes. These forces still operate. Why should such forces disappear just because we humans live on Earth?

The fundamental questions remain unanswered. A change of 1 per cent in cloudiness can account for all changes measured during the past 150 years, yet cloud measurements are highly inaccurate. Why is the role of clouds ignored? Why is the main greenhouse gas (water vapour) ignored? The limitation of temperature in hot climates is evaporation yet this ignored in catastrophist models.

Why are balloon and satellite measurements showing cooling ignored yet unreliable thermometer measurements used? Is the increase in atmospheric CO2 really due to human activities?

Ice cores show CO2 increases some 800 years after temperature increase so why can't an increase in CO2 today be due to the medieval warming (900-1300)?

If increased concentrations of CO_2 increase temperature, why have there been coolings during the past 150 years?

Some 85 per cent of volcanoes are unseen and unmeasured yet these heat the oceans and add monstrous amounts of CO_2 to the oceans. Why have these been ignored? Why have there been five significant ice ages when CO_2 was higher than now? Why were warmings in Minoan, Roman and medieval times natural, yet a smaller warming at the end of the 20th century was due to human activities? If climate changed at the end of the Little Ice Age (c.1850), is it unusual for warming to follow?

Computer models using the past 150 years of measurements have been used to predict climate for the next few centuries. Why have these models not been run backwards to validate known climate changes?

I would bet the farm that by running these models backwards, El Nino events and volcanoes such as Krakatoa (1883, 535), Rabaul (536) and Tambora (1815) could not be validated.

In my book, I correctly predicted the response. The science would not be discussed, there would be academic nit-picking and there would be vitriolic ad hominem attacks by pompous academics out of contact with the community.

Comments by critics suggest that few have actually read the book and every time there was a savage public personal attack, book sales rose. A political blog site could not believe that such a book was selling so well and suggested that my publisher, Connor Court, was a front for the mining or pastoral industry.

This book has struck a nerve. Although accidentally timely, there are a large number of punters who object to being treated dismissively as stupid, who do not like being told what to think, who value independence, who resile from personal attacks and have life experiences very different from the urban environmental atheists attempting to impose a new fundamentalist religion.

Green politics have taken the place of failed socialism and Western Christianity and impose fear, guilt, penance and indulgences on to a society with little scientific literacy. We are now reaping the rewards of politicising science and dumbing down the education system. If book sales, public meetings, book launches, email and phone messages are any indication, there is a large body of disenfranchised folk out there who feel helpless. I have shown that the emperor has no clothes. This is why the attacks are so vitriolic.

Ian Plimer is emeritus professor of earth sciences at the University of Melbourne. His book Heaven and Earth is published by Connor Court."
Ian Plimer,
The Australian, May 29th 2009
Vitriolic climate in academic hothouse

Recommend checking out his book online...

Heaven and Earth: Global Warming, the Missing Science

www.amazon.com/Heaven-Earth-Warming-Missing-Science/dp/1589794729

CHAPTER 9

Summary

"Those who don't know history are doomed to repeat it." - Edmund Burke

What have I learned from organizing this book and reading over 100+ articles from four decades of research and commentary on the subject of climate change? Personally I feel the quote below explains my take away better than I could.

"It is more difficult to undermine faith than knowledge, love succumbs to change less than to respect, hatred is more durable than aversion, and at all times the driving force of the most important changes in this world has been found less in a scientific knowledge animating the masses, but rather in a fanaticism dominating them and in a hysteria which drove them forward."
Adolf Hitler, Mein Kampf

Hitler was a mad man who knew how to radicalize a nation against the world and now it seems this philosophy is being used to radicalize the world behind socialist ideals by using climate as the catalyst. Sure many readers might disagree but the evidence within this book clearly proves my conclusion.

Below are a few questions I leave for you to contemplate.

Who had the idea to start Earth Day? A Democrat by the name of Gaylord Nelson, then a U.S. Senator from Wisconsin

Who was the global warming alarmist who created the movie an Inconvenient Truth which was proven in court to have nine factual errors? A Democrat by the name of Al Gore, former Vice President of the United Stated of America

Who signed the Paris Climate Accord without Congressional Approval? A Democrat by the name of Barack Obama, then the 44th President of the United Stated of America

Here is an excerpt from the Democrats.org website which states, *"Democrats believe that climate change poses a real and urgent threat to our economy, our national security, and our children's health and futures…"*. Their website also states, *"Democrats are committed to curbing the effects of climate change…"* which means the Democrats would continue throwing Government money, the peoples money, at Policy which may or may not actually help deal with figuring out the cause and severity of the human impact on climate change.

It was as recent as May 2017 when former President Obama stated;

"During the course of my presidency, I made climate change a top priority because I believe that of all the challenges that we face, this is the one that will define the contours of this century, more dramatically perhaps than any other."
Nic Robertson, Jacopo Prisco and Angela Dewan
CNN, May 9, 2017
Obama defends Paris climate accord as Trump mulls ditching it

Climate Change was a bigger priority than keeping Americans safe? A bigger priority than our economy? A bigger priority than everything else Country related?

President Obama was the first President to nearly double the National debt from 10 trillion dollars to over 19 trillion dollars. Where did all those trillions of dollars go? We know the Paris Accord would have been a massive redistribution of American wealth if fully realized. Why would a President sign an international agreement, without congressional approval, which would effectively steal money from the people he was sworn to serve?

There are so many questions I have this summary could go on forever so I'll end with this quote.

"It is part of a great leader's genius to make even widely separated adversaries appear as if they belonged to but one category, because among weakly and undecided characters the recognition of various enemies all too easily marks the beginning of doubt of one's own rightness."
Adolph Hitler, Mein Kampf

Hitler's most important individual contribution to the theory and practice of National Socialism was his deep understanding of mass psychology and mass propaganda. He stressed the fact that all propaganda must hold its intellectual level at the capacity of the least intelligent of those at whom it is directed and that its truthfulness is much less important than its success.

I'd like to thank the Editors of Encyclopedia Britannica for the last paragraph and also thank all of you for caring enough to try to decipher the truth behind the climate change craze.

CHAPTER 10

Globalcide

"If liberty means anything at all, it means the right to tell people what they do not want to hear." - George Orwell

Definition of *globalcide*

1. When the environment is used as a means to isolate a people for extermination

The Socialists/Progressives will attempt to use Globalcide to cleanse the world of so called Climate Deniers.

"Even death is not to be feared by one who has lived wisely." - Buddha

"The hard and stiff are death's companions. The soft and weak are life's companions." - Tao Te Ching

"What shall we then say to these things? If God be for us, who can be against us?" — King James Version, Romans 8:31

CHAPTER *11*

Theme Song

"Some of us think that we know everything and more,
And so we have the right to break all the rules." - Clinton Fearon

Clinton Fearon, Goodness 2014
Song: Wi No Know It

www.youtube.com/watch?v=czkMMm4Fj60

CHAPTER 12

Von Knowledge Quotes

"Share the Knowledge, Square the Knowledge."- Von Knowledge

1. *"A compromised media becomes a polarized media."*
   Von Knowledge

2. *"The professor who becomes an alarmist without all the facts does a disservice to his profession."*
   Von Knowledge

3. *"Government funding makes some climate scientists chummy."*
   Von Knowledge

4. *"Global cooling heated up the climate industry, global warming got the media fired up."*
   Von Knowledge

5. *"Doomsday predictions are like snow in the summer."*
   Von Knowledge

6. *"Riddle me this, riddle me that, the climate scientist without all the facts, not only breaks the industry's back but exposes his lack... of knowledge, reason, and tact."*
   Von Knowledge

7. *"Socialism is but another name for Totalitarianism."*
   Von Knowledge

Additional Quotes

"As a climate researcher, I am increasingly convinced that most of our recent global warming has been natural, not manmade."
Roy W. Spencer, Meteorologist
Principal Research Scientist at the University of Alabama in Huntsville

"Given that the evidence strongly implies that anthropogenic warming has been greatly exaggerated, the basis for alarm due to such warming is similarly diminished."
Richard S. Lindzen, Atmospheric Physicist

"I find no compelling reason to believe that the earth will necessarily experience any global warming as a consequence of the ongoing rise in the atmosphere's carbon dioxide concentration."
Sherwood B. Idso, President
Center for the Study of Carbon Dioxide and Global Change

"A number of studies point to sources other than greenhouse gases as explanations for the modest warming trend of the late 20th century."
Patrick J. Michaels, Climatologist
Senior fellow at the Cato Institute

"I'm sure the majority (but not all) of my IPCC colleagues cringe when I say this, but I see neither the developing catastrophe nor the smoking gun proving that human activity is to blame for most of the warming we see."
John R. Christy, Climate Scientist
University of Alabama in Huntsville

"We see no evidence in the climate record that the increase in CO2, which is real, has any appreciable effect on the global temperature."
S. Fred Singer, Physicist and Emeritus Professor
Environmental Science at the University of Virginia

CHAPTER 13

Only Time will Tell

"He who controls the past controls the future. He who controls the present controls the past." - George Orwell

These scientists have said that it is not possible to project global climate accurately enough to justify the ranges projected for temperature and sea-level rise over the next century.

David Bellamy, Botanist

Lennart Bengtsson, Meteorologist

Piers Corbyn, astrophysicist, owner of the business WeatherAction which makes weather forecasts

Judith Curry, Professor and former chair of the School of Earth and Atmospheric Sciences at the Georgia Institute of Technology

Freeman Dyson, professor emeritus of the School of Natural Sciences, Institute for Advanced Study; Fellow of the Royal Society

Ivar Giaever, Norwegian–American physicist and Nobel laureate in physics (1973

Steven E. Koonin, theoretical physicist and director of the Center for Urban Science and Progress at New York University

Richard Lindzen, Atmospheric Physicist

Alfred P. Sloan emeritus professor of atmospheric science at the Massachusetts Institute of Technology and member of the National Academy of Sciences

Craig Loehle, ecologist and chief scientist at the National Council for Air and Stream Improvement

Ross McKitrick, Professor of Economics and CBE Chair in Sustainable Commerce, University of Guelph

Patrick Moore, former president of Greenpeace Canada

Nils-Axel Mörner, retired head of the Paleogeophysics and Geodynamics Department at Stockholm University, former chairman of the INQUA Commission on Sea Level Changes and Coastal Evolution (1999–2003)

Garth Paltridge, retired chief research scientist, CSIRO Division of Atmospheric Research and retired director of the Institute of the Antarctic Cooperative Research Centre, visiting fellow Australian National University

Roger A. Pielke, Jr., professor of environmental studies at the Center for Science and Technology Policy Research at the University of Colorado at Boulder

Tom Quirk, corporate director of biotech companies and former board member of the Institute of Public Affairs, an Australian conservative think-tank

Denis Rancourt, former professor of physics at University of Ottawa, research scientist in condensed matter physics, and in environmental and soil science

Harrison Schmitt, geologist, Apollo 17 Astronaut, former U.S. Senator.[62]

Peter Stilbs, professor of physical chemistry at Royal Institute of Technology, Stockholm

Philip Stott, professor emeritus of biogeography at the University of London

Hendrik Tennekes, retired director of research, Royal Netherlands Meteorological Institute

Anastasios Tsonis, distinguished professor of atmospheric science at the University of Wisconsin-Milwaukee

Fritz Vahrenholt, German politician and energy executive with a doctorate in chemistry

These scientists have said that the observed warming is more likely to be attributable to natural causes than to human activities.

Khabibullo Abdusamatov, astrophysicist at Pulkovo Observatory of the Russian Academy of Sciences

Sallie Baliunas, retired astrophysicist, Harvard-Smithsonian Center for Astrophysics

Timothy Ball, historical climatologist, and retired professor of geography at the University of Winnipeg

Ian Clark, hydrogeologist, professor, Department of Earth Sciences, University of Ottawa

Chris de Freitas, associate professor, School of Geography, Geology and Environmental Science, University of Auckland

David Douglass, solid-state physicist, professor, Department of Physics and Astronomy, University of Rochester

Don Easterbrook, emeritus professor of geology, Western Washington University

William Happer, physicist specializing in optics and spectroscopy; emeritus professor, Princeton University

Ole Humlum, professor of geology at the University of Oslo

Wibjörn Karlén, professor emeritus of geography and geology at the University of Stockholm

William Kininmonth, meteorologist, former Australian delegate to World Meteorological Organization Commission for Climatology

David Legates, associate professor of geography and director of the Center for Climatic Research, University of Delaware

Anthony Lupo, professor of atmospheric science at the University of Missouri

Tad Murty, oceanographer; adjunct professor, Departments of Civil Engineering and Earth Sciences, University of Ottawa

Tim Patterson, paleoclimatologist and professor of geology at Carleton University in Canada

Ian Plimer, professor emeritus of mining geology, the University of Adelaide.

Arthur B. Robinson, American politician, biochemist and former faculty member at the University of California, San Diego

Murry Salby, atmospheric scientist, former professor at Macquarie University and University of Colorado

Nicola Scafetta, research scientist in the physics department at Duke University

Tom Segalstad, geologist; associate professor at University of Oslo

Nir Shaviv, professor of physics focusing on astrophysics and climate science at the Hebrew University of Jerusalem

Fred Singer, professor emeritus of environmental sciences at the University of Virginia

Willie Soon, astrophysicist, Harvard-Smithsonian Center for Astrophysics

Roy Spencer, meteorologist; principal research scientist, University of Alabama in Huntsville

Henrik Svensmark, physicist, Danish National Space Center

George H. Taylor, retired director of the Oregon Climate Service at Oregon State University

Jan Veizer, environmental geochemist, professor emeritus from University of Ottawa

These scientists have said that no principal cause can be ascribed to the observed rising temperatures, whether man-made or natural.

Syun-Ichi Akasofu, retired professor of geophysics and founding director of the International Arctic Research Center of the University of Alaska Fairbanks

Claude Allègre, French politician; geochemist, emeritus professor at Institute of Geophysics (Paris)

Robert Balling, a professor of geography at Arizona State University

Pål Brekke, solar astrophycisist, senior advisor Norwegian Space Centre

John Christy, professor of atmospheric science and director of the Earth System Science Center at the University of Alabama in Huntsville, contributor to several IPCC reports

Petr Chylek, space and remote sensing sciences researcher, Los Alamos National Laboratory

David Deming, geology professor at the University of Oklahoma

Stanley B. Goldenberg a meteorologist with NOAA/AOML's Hurricane Research Division

Vincent R. Gray, New Zealand physical chemist with expertise in coal ashes

Keith E. Idso, botanist, former adjunct professor of biology at Maricopa County Community College District and the vice president of the Center for the Study of Carbon Dioxide and Global Change

Antonino Zichichi, emeritus professor of nuclear physics at the University of Bologna and president of the World Federation of Scientists

Kary Mullis, 1993 Nobel Laureate in Chemistry

Scientists arguing that global warming will have few negative consequences. These scientists have said that projected rising temperatures will be of little impact or a net positive for society or the environment.

Indur M. Goklany, science and technology policy analyst for the United States Department of the Interior

Craig D. Idso, faculty researcher, Office of Climatology, Arizona State University and founder of the Center for the Study of Carbon Dioxide and Global Change

Sherwood B. Idso, former research physicist, USDA Water Conservation Laboratory, and adjunct professor, Arizona State University

Patrick Michaels, senior fellow at the Cato Institute and retired research professor of environmental science at the University of Virginia

The last six pages were taken from Wikipedia's "List of scientists opposing the mainstream scientific assessment of global warming" page.

https://en.wikipedia.org/wiki/List_of_scientists_opposing_the_mainstream_scientific_assessment_of_global_warming

I'd like to thank these courageous scientists for standing up for what they believe in and challenging those who attempt to control the climate narrative. Without you staying true to your core principles and values the public would never have a chance to fully understand the scope, depth, mystery of our planets climate history.

Please share this book with your friends and families, you're Freedom just might depend on it!

"Sharing is caring, stealing from others is not."
Von Knowledge